Landscape Papers Edgar Anderson
Edited, with an Introduction, by Bob Callahan

Turtle Island Foundation
Berkeley
1976

QH
81
A52
X
——
T6451

LANDSCAPE PAPERS are published by Turtle Island for the Nezahaulcoyotl Historical Society, a non-profit corporation engaged in the study of the history and literature of the New World. Copyright © 1976 by Landscape Magazine. Special acknowledgment is made to J. B. Jackson, former Landscape editor, and to Blair Boyd, current editor.

Library of Congress Catalog Card Number 73-43625
ISBN 0-913666-11-4

All rights reserved. Printed in the U.S.A. For further information address: Nezahaulcoyotl Historical Society, 2845 Buena Vista Way, Berkeley, CA 94708.

CONTENTS

Introduction	9
Spring and Summer and the Green Lull Between	13
The Considered Landscape	21
Horse and Buggy Countryside	24
College and the Experience of Nature	28
The Cornbelt Farmer and the Cornbelt Landscape	33
Confederate Violets	36
Maize of the Southwest	43
Sunflowers in the West	48
Anatolian Mystery	52
The City is a Garden	58
Autumn on the Peninsula	64
Reflections on Certain Honduran Gardens	68
Living with a Gingko	74
The Country in the City	78
Islands of Tension	85
The City Watcher	90

Introduction

The idea for this gathering of Edgar Anderson's *Landscape Papers* finds its source, easily enough, in the epilogue attached to later editions of Anderson's own earlier book of magic, *Plants, Man and Life*. "Shortly after *Plants, Man and Life* appeared, it was reviewed in the journal *Landscape*," Anderson writes. "This led me to J. B. Jackson, the able editor of that remarkable publication. With piquant combination of sharp criticism and flattering appreciation, he charmed out of me a series of short essays. One of these essays takes up in more detail the "dumpheap theory" of the origin of cultivated plants. Several touch in one way or another on the acceptance of cities as places to live right in the middle of; an attitude that is part of Mexico's Spanish heritage. Three of the essays describe how I learned enough from my Mexican neighbors to have lived serenely years later in a big, moderately priced St. Louis apartment hotel for six months, as a naturalist happy in learning from our apartment windows: new things about bird life in the big city, the serpentine course (unparalleled in its loopings) of the Missouri Pacific railroad as it starts southward, the progress of small thunderheads across the city on a day of little thunder showers, the dynamics of winter sunsets when the sun is so low in the sky that the observer needs to be well over a hundred feet above the ground level to learn very much, and the heights and extents of autumnal morning fogs. I was again spurred on to study and to teach the natural history of cities along with that of seashores and spring woodlands. Some of those who liked this book have found deeper satisfactions in these *Landscape* papers."

•

Many impressions surface upon reading, and re-reading these Anderson papers. I am struck over and again by the detail of the man's knowledge, by the elegance of his prose—"It's the utter transparency of Hazlitt that has been my model and should be a model for us all," he wrote, and perhaps most importantly of all, by a certain reflective quality of attention and observation deeply rooted, I suspect, in his own Quaker way of life.

"This curious world is unendingly fascinating," Edgar Anderson's friend Carl Sauer, once said. Edgar Anderson, I take it, would have been the first to agree. We have had two great American Andersons in this century, both born and raised on the transplanted cornbelt mysticism of our own Middle Border: one, his name was Sherwood, sung *Mid-American Chants* and created a *Winesburg, Ohio*; and this other, this Edgar, a botanist, sought after new historical directions. They are related not in blood, but by spirit—the new life that follows when the spirit of wonder, touch and concern is again about the land. It is a thrill to be associated with this book.

                    Bob Callahan
                    Berkeley, 1976

LANDSCAPE PAPERS

Spring and Summer and the Green Lull Between

Sometimes the American poet lets us down. He lives and writes in America but in his mind's eye he looks upon an English rather than an American landscape. Look about you just as spring is passing into summer; look with a clear eye and a critical mind and see if you find the kind of a June the average American poet has been singing of. Don't just sit in your garden. Drive through the country with your eye on the landscape. I can tell you what it is like in the southern half of the corn belt and it is not the traditional June of the facile English lyric or the sonorous English sonnet. It is a green world, in good years a lush green one, green along the roadsides, green in the woodlands, hot steaming green in the heavy vegetation along the rivers. It is the "green lull." The burst of spring bloom is over; the redbud and the dogwood and wild plum no longer stand out along the wood edges; the trumpet vines, the button bushes, the big saucer-shaped blooms of the marshmallow have not yet made their appearance. If it were not for the European plants we have brought along with us to our gardens and hayfields there would scarcely be a blossom.

Having had something to do with the management of a

wild flower reservation in eastern Missouri, for several years I kept a careful record of the conspicuous wild flowers in bloom each week. It started out by being a simple weekly check-up for practical purposes; it has grown into a good deal more than that. The data have been used to study blooming seasons, to make exact comparisons of meadow bloom versus woodland bloom. It is fifteen years since I first analyzed the data and they still give me a good deal to think about. It was these data that first called my attention forcefully to what I have come to think of as the green lull.

We began our study by choosing around a hundred species in the local flora which would be of particular interest to the public visiting a wild flower reservation, species like the wild sweet William and the spring beauty which made masses of color in the landscape or others like lady's tresses orchids so charming that they are of general interest even though comparatively inconspicuous. The list was checked over every weekend throughout the year and each species was listed as out of flower, or coming into bloom, or in full bloom, or going out of bloom. The scoring in each case was by the general effect in the landscape. The redbud was recorded as coming into bloom when it first began to show purple on the hillside, it was listed as going out of bloom when it gave a patchy effect in the landscape. If we are trying to study the progress of the season, to chart nature's tides of bloom, then this simple record by landscape effect is better than it looks at first sight. The few exceptional flowers which come out ahead of time do not muddy up the record, those which linger on one or two after all the rest are past are similarly ignored and we have a record which is a really efficient reflection of the season's progress from week to week.

By throwing all the data together in a sort of index we can produce a curve analogous to those constructed for charting the rise and fall of business week by week. Every species of full bloom we counted as one point on the index, species coming into flower at $1/3$ of a point, and those passing out of flower were scored at $2/3$ of a point; the scores were totted up for each week. It makes an interesting curve. It rises slowly in late winter with the little things which creep into bloom during warm spells, the "pepper-and-salt," the hepatices, the early bluets. Then it rises rapidly to a peak and as rapidly falls. The end of April finds over twenty per cent of our sample in bloom but by early June (the month the poets sing of) there are no

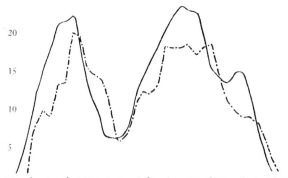

Number of common wild flowers in bloom for two successive years from March 1 to November 15.

more kinds of wild flowers in bloom than there were the middle of March. This is the green lull, the pause between spring and summer bloom. In the latitude of St. Louis it is a definite break in the wild flower parade. To the south of us it is even more prolonged. As one goes northward it becomes less and less marked. In southern Michigan it is merely a dip between the peak of spring and summer flowering. Farther north, I imagine, it would disappear altogether.

Summer bloom has begun by the first of July. It rises slowly past mid-summer and then may sport a little autumnal peak when the fall composites, the asters, goldenrods and such like plants come hurrying into bloom just before the frosts.

With the curve before us we see the tide of seasonal bloom more clearly than we could before. The whole season is there at a glance. There is the rapid surge of spring, its ebbing, equally rapid, to the June lull. There is the long plateau of summer bloom and the sudden glory of the autumnal composites just before the killing frosts. So precise is the curve that it becomes more than just a record, a precise aid to the memory. It can be used for analysis. We can compare the bloom of one season with that of another or chart the seasonal differences exactly from place to place. With these exact comparisons before us, we may hope not only to describe the process but to figure out some of the controlling factors.

A good place to begin is to compare the curves for several autumns. If we consider how strikingly rainfall and temperature can vary from year to year in mid-continental Missouri, the remarkable thing about these curves is their similarity from year to year. The parade of bloom is much more alike from year to year than is the parade of weather. Early killing frost or late killing frost make little or no difference. Autumnal rain or autumnal drought may effect the amount of bloom; they have almost no effect on its timing. Except for a few stragglers the brilliant fall blooming season comes rapidly to an end with great regularity. The curve starts down in September; by late October it has reached the bottom. The descent of the spring curve is quite as regular but not its ascent. In a cold late spring, the curve may be two weeks or more later than in a sunnier one, as it rises to a peak. Not so its descent. Late spring or early spring has less effect on this part of the curve. The position of the green lull, like cessation of autumnal bloom, is virtually independent of year to year differences in temperature and rainfall.

The responsible factor is not far to seek. It is very probably the length-of-day. Many plants are extraordinarily sensitive to day length, and their vital processes, their coming into bloom, their ripening of seeds, their going out of bloom are keyed to day length. Get seeds of corn or beans from Central America and try growing them in your own garden. Most of these tropical varieties, if you plant them in spring, will grow and grow, waiting for the proper day length which initiates bloom. Corn which was only head high in Guatemala will, when planted in your garden, grow up and up through the long days of summer, with never a sign of tassel or silk. Finally as fall comes on and the shortening days reach a length comparable with that of Guatemala's growing season, out pop the tassels and the silks, high over head. Or try the opposite experiment. Take a little flint corn from northern Canada, ordinarily shoulder high and grow it in Iowa. Baffled by the differences in day length, it comes into tassel when it is scarcely waist high and heroically begins to make an ear before it is mature enough to support one. Plant it still farther south and it will tassel out when it is scarcely knee high and produce a pitiful attempt at an ear close to the surface of the ground.

If this evidence fails to convince you take a black box, or an old oil barrel, and give a patch of wild flowers autumnal day length in June. Put the box over them every night when you come home from work and take it off at breakfast time. In a few weeks at most, some of the asters and goldenrods will have adjusted themselves to a short day economy and will set flowerbuds all over the top of the plants as if September were already here.

Day length is not the only control which nature uses in keeping bloom and seed-set in time with the march of the seasons, but is the main one for many plants, and its independence of differences in rainfall and temperature give our blooming seasons a regularity they would otherwise lack. The dramatic rise of flowering from early spring to its height is more completely understood if we compare the blooming curves for different habits. Missouri is close enough to the Southwest so that we have rocky cliffs and stony barrens known locally as glades. If we compare the blooming curves for glade and cliff with those from spring woodlands the difference is revealing. For glade and cliff the curve rises slowly; the spring curve fades into summer with scarcely a dip. For the woodlands it rises sharply and as sharply falls to zero, and close to zero it remains all summer.

Here we are dealing with two effects of season, the effect in any one year and the effect in terms of evolution. Our spring woodlands and our glades have been with us for a long time; habitats similar to each have been here even longer. Natural selection has given us the kinds of spring wild flowers which fit into these two habitats. Take the woodlands. They are warm and sunny in early spring. The rich woods dirt is a good medium for perennials and the soil is moist. When the trees burst into leaf, all this is changed. There is little or no sunshine on the forest floor and the tree roots are busy pumping incredible amounts of water up to the trunks and branches. Except immediately after a rain the upper few inches of the soil are drier than it should be in a good perennial border. The glades, by contrast, are much the same spring and summer. No need there for a species to bloom in April or early May. Late May or June would serve as well. In the woodland natural selection has been a driving force not only in this season but for thousands of seasons in the life of the species which flourish there. The woodlands of early spring were ideal for flowering perennials, the summer woodlands were not. There has therefore been pressure, generation after generation, to evolve those flowers which would come into bloom rapidly, go to seed and die down by early summer. There has been every inducement for the evolution of wild flowers which could fit in between the last killing frost of spring and the time when there is a solid canopy of leaves overhead. That burst of spring bloom was bred for our spring woodlands and day length is one of the main controls nature has used in forcing it to perform within these strict limits.

The Semi-tropical Summer Flowers

It is interesting to contrast the flowers of the spring woodlands with our summer wild flowers. Those of spring come quickly and go quickly. If you visit the same woodland on successive weekends in the springtime, you see its aspect change dramatically from week to week. On your last visit the low woods were blue with phlox; this week the phlox are passing out of the picture; next week phacelia will be in full bloom and there will be sheets of foamy bluish white replacing the soft clear blue. Trillium, dutchman's breeches, phlox, few of these are effective in the landscape for more than three weeks. Not so the flowers of summer. The button bush with its fragrant white balls of bloom is in flower for six weeks; the trumpet vine, beloved of the hummingbirds, come slowly into flower in late June but does not go completely out of bloom until well into September. We have nothing in the spring woodlands which anywhere near matches this performance.

This again, I believe, is a by-product of evolutionary history. Most of the spring flowers are not only strictly limited to the temperate zone, the genera from which they sprang are northern in their affinities. Our summer bloomers are immigrants from the tropics. Take any of our spring woodland wild flowers and chart their known distribution on a map. Characteristically they are centered upon the interior plateaus of the eastern United States. They seldom reach the Gulf Coast or the Carolina coastal plain. None of them are in Peninsular Florida. Contrast them with most of our summer bloomers. Trumpet vine, partridge pea, and button bush reach the coastal plain and extend down into Florida.

The contrast is even more striking if we examine all the close relatives of these same plants. The spring flowers belong to north temperate genera. Trilliums are found across the Continent and on into Asia but always in the temperate zone. There are trilliums in Alaska and trilliums in North Japan. Our adder's-tongues and shooting stars have relatives in the west but they are all northern in their likes and dislikes and they run south, if at all, only along the mountains.

Not so the summer bloom. Our trumpet vine has the northern-most distribution of anything in its entire family, which is largely made up of the flowering trees and lianas. It is the only species in all this assemblage whose main distribution is within the temperate zone though even it grows partly within the sub-tropics. The trumpet vine is tropical even in its effect in the landscape. Study it where it grows in native profusion along river margins. It climbs high up into the tallest trees along the river bank and where it can get out to the sun it breaks out in handsome trusses of flaming orange red flowers. To one who has been in the tropics it creates exactly the same effect as its liana relatives of Brazil and Venezuela. As each flower reaches the end of its blooming span, the corolla, still quite fresh, falls off in one piece and lies on the damp soil of the river bank or floats down the river, just as it would in the tropics. And if you come down the same stretch of river a fortnight later, the same vines will probably be blossoming in the same trees and today's crop of freshly fallen blooms will be lying on the damp soil or floating down the river. So it goes for most of our other summer bloomers. The button bush is a relative of the coffee plant and is the only rubiaceous shrub to reach the temperate zone. Bumelia, a strange little tree of our Missouri glades which blooms in the hottest part of midsummer, is the northernmost species, of the northern-most genus, of an entire order of tropical plants. The partridge pea which produces sheets of yellow along our roadsides in July and August belongs to a group of plants otherwise almost unknown in the temperate zone but common throughout the world in the tropics and sub-tropics. Our summer bloom is made up almost entirely of immigrants from the tropics and the sub-tropics which have somehow become adapted to a dormant season which is not only deficient in rainfall (as in the tropics) but is also extremely cold.

Truth and Poetry in the Landscape

Such in broad outline, is the progress of bloom throughout the season in a good part of eastern North America. Similar curves could easily be established for other parts of the country, and the interpretation of the main factors in their ebbings and flowings would be as simple. In dull factual detail blooming season is a subject for the biologist; in these broad aspects it deserves more general attention. The general facts about it should be known to geographers and those who deal theoretically or practically with problems of the landscape. It is quite as much a field for the poet and the essayist, for natural philosophers in the strictest sense of the term. Few scientists today are whole men; not enough poets and philosophers are grounded in a sensitive awareness of the actual world in which they live. Crescent moons, in too much poetry, rise in the East at sunset time and June is the month of flowers. Eventually these simple facts about our American seasons should pass into folk wisdom. Then American poets sensitive to the peculiarities of our American environment would transmute them into poetry as apposite to the place in which it was written as Keats' Endymion or the Book of Job were to their own environments.

The Considered Landscape

When we consider a landscape what *are* we considering? Is it just what we see or is it something more—if so, what is that something more? What we see is a view, most certainly. When we talk about landscape, when we try to have a meeting of minds as to its various problems, there is more than the view itself. We are *contemplating* what is before us. The eye is seeing and the mind is perceiving. What we think, what we ask, what we investigate will depend upon how rich is the experience brought to bear on that contemplation. It is not only what we see, it is also what we see *in* it.

As a simple illustration, let me take the landscape of the Conewango Valley in western New York State and suggest something of its effect upon me at the age of ten, of thirty and of sixty. In childhood it was to me the broad flat valley which separated the Cataraugus County hills of one grandparental home from the Chatauqua County hills where my other grandfather lived, and my uncles and cousins. Coming back to it from the general flatness of southern Michigan, I was intrigued by the gray-blue hill masses which piled up against the sky, hills set upon hills in terms of our Michigan landscapes.

I was intrigued by the differences though I did not then realize that these rolling New York hills are rock-ribbed under a thin overlay of glacial deposits and that our Michigan hills had no real skeletons, being heterogeneous giant moraines dumped by the last vigorous pushes of a continental glacier.

When I came back again to the Conewango Valley in my thirties I saw all this and thought about it. I realized that those little flat fields all in a line, which edge the valley so prettily, are the terraces of an early post-glacial lake which once covered all the lowlands, like the Finger Lakes of sister valleys to the eastward. I knew the secret at last of the strange views to the south which had so intrigued me in my later boyhood, row after row of flattish-topped high hills all breaking in regular sequence. This rhythmic pattern of the Allegheny and its tributaries, the never-never land of my boyhood dreams, those strangely different and serried hill systems that I could see to the south from the highest hills on the clearest days, were the unglaciated Allegheny plateau, just beyond easy reach in horse-and-buggy days.

In my thirties, the automobile was putting these Pennsylvania landscapes within an afternoon's drive, but it was having other and far-reaching effects in the Conewango Valley and among the nearby hills. One could still go in to Mr. Bagg's meadow for a picnic but he had painted a sign on the gate: "Yes, you can come in if you want to, but for God's sake close the gate!" The smaller hamlets were now dying off as centers of trade; the little stores of Leon had mostly closed their doors and even those of larger Ellington did not look so prosperous. The daisy-dotted upland meadows of my boyhood still delighted me, though I now realized that these hill pastures were no longer productive. The flowers and the sweet-scented fern were mute evidence of why the barns and houses were not freshly painted. Young people were being drained off to the towns and cities; dairy farming with its onerous chores had no longer a strong appeal to younger generations; the whole countryside was slowly draining downhill.

What a delight to come back at sixty and see a full tide setting in, and in the other direction: electricity and invention had erased the isolation and the drudgery of the farm. Grassland farming was coming into its own. Even on upland pastures there were more cows grazing in one field than in my grandfather's day could have been supported on an entire farm. The daisies and buttercups were not conspicuous in the pastures with the cows. My farmer cousins were no longer faintly amused at my botanical preoccupation with the vegetation under foot; instead they eagerly joined in technical discussions of "birdsfoot trefoil" and improved strains of clover. The increasing political might of farmers was evident in the attractive buildings around the town square in Ellington; a new kind of dispersed urbanism was on the doorstep. I needed my wisest friends in economics and sociology to help me interpret the dynamics of this changing landscape before me.

Horse and Buggy Countryside

It is gone and gone forever; like youth it was not fully appreciated until it was over. I do not wish it back; it was for me an inconvenient and a confining landscape, but neither should it be forgotten. Protest wells up that its sights and sounds and smells are lost forever. That it should go without a record of what it was like, some indication of what it meant to those who lived in it, seems as out of place as the passing of a great statesman without a word as to his services to the country. Miss Cather has written very beautifully about the dust of country roads; there are phrases here and there in Hamlin Garland; but I do not remember having found anywhere an attempt to record the impact of the horse-and-buggy landscape upon those who lived and moved in it.

Impact it had and a very sharp and constant one. You couldn't be unconscious of the terrain over which you were traveling. The road recurringly thrust itself upon your senses; there were vibrations, sounds, and smells. The wheels commented almost constantly about the roadway. There was a soft even purr as they pulled slowly through the sand, anguished crunches in sliding over the edges of big stones, a taut, almost musical, vibration when they whirled rapidly along on good gravel. There were waits for the horses to rest when the grade was steep. Swamp roads were an adventure; plank bridges over the ditches, heavy new gravel that almost stalled the wheels, miry spots where the buggy lurched unevenly and, sometimes, bumpety-bump stretches of corduroy when the logs and brush which held the road up out of the swamp were still close to the surface.

Such roads had their own standards of courtesy. I remember the story told me by the first man to drive an automobile on the back roads around Fairlee, Vermont. He came up behind a farmer with a load of hay midway through the balsam swamp. It was a narrow road; wagons could just barely pass if they met head on, and a faster team, coming up behind a slower one, stayed behind until the swamp had been crossed. My friend tooted his horn vigorously, but there was no response from the hay wagon. Finally he stopped, stuck out his head and yelled, "Hey, don't you hear me up there?" The farmer reined in his horses, stood up, looked down with great dignity and said, "Oh yes, I hear you all right," and drove ahead slowly and unconcernedly to the end of the swamp where he pulled off to the side as he would have for any other vehicle which had come up behind him. "He was right and I was wrong," said my friend years later, "but I didn't realize that at the time."

In those days differences in vegetation between a main highway and a side road were even greater than now. There weren't many state highways but the best of them stretched straight ahead with mathematical perfection, even though just of gravel. These rights-of-way were usually wide and from the road edge to the boundary fences were mowed pretty much as now, though road building had not been preceded by earth moving. As a result, there were more native plants in among the grass and fewer of those cosmopolitan tramps which take so readily to disturbed habitats. But the side roads, those fascinating side roads, they are mostly gone and there is nothing in the modern road system to compare with them. The wheel track wound here and there, depending upon the slope and the vegetation and the character of the land. Sometimes it was straight for a little way; frequently it wobbled. There was often grass in the roadway outside the actual wheel tracks; shrubs like sumac and elderberry pressed so close to the road that you could smell them as you drove by and children snatched at the flowers. Accommodating drivers of the local stage-line learned to snip off small twigs with a snap of the buggy whip and present them to lady passengers.

The biggest difference of all was to the ear; you then traveled through the sounds of the country quite as much as through its sights. On clear days in spring there was the incessant calling of the meadow lark, and in the heavy heat of early summer you heard repeatedly the scolding call of the Maryland yellowthroat coming out of the dusty bushes along the roadside.

Even more characteristic than these sharp sounds were the soft and gentle ones that are scarcely heard any more, anywhere. So crisscrossed is our modern landscape with the pounding of Diesel trucks and farm tractors that over square miles of countryside, even when standing quietly in fields or woods, there is enough interfering of mechanical noise to keep one from hearing the softest of the cricket trillings on autumn evenings, the gentlest of breezes in the long-needled pines, or that magic sound of the wind in the young corn. When the corn is about waist-high, the leaves are not yet at all stiff. They bend and twist in the wind and touch each other lightly. If there are no other noises, you can hear the slightest breezes rippling here and there through the cornfield. There is no other sound like it; on a clear, quiet day in early summer when the wind rises and then dies away again, this mystical rustling (like thousands of silk petticoats just barely being stirred to sound by gentle movements) scurries back and forth over the landscape.

Or the onset of snow; who hears this now? The first flakes as they start to fall hit the dry leaves and the dead grasses along the roadsides with tiny, sharp noises, like the ghosts of mice in baskets of vanished waste paper. They begin as sharp, separate rustles, becoming more and more blended as the snow falls in earnest, cushioning the cart wheels into silence. It was beautiful to hear, but to the alert traveler it was a sign of trouble ahead, and if one had far to go, it had ominous associations.

No, the horse-and-buggy landscape was not just delight. The finest days were finer; one got everything from them that could be collected from a fine day, but the harsh ones were harsher. There was misery as well. I remember the farm families driving home in the early dusk of winter afternoons. I used to meet them in the swamp road, as I trudged home after a Saturday tramp through marshlands frozen so solid that one could easily explore the quagmires which were impossible in the summertime. The farmers would have done their weekly shopping and the wagons jolted noisily over the frozen ruts. The driver usually sat alone up front, hunched down against the bitter wind. His wife and children, together with the purchases for the week, were down in the straw that had been spread over the wagon bed, swathed in heavy horse blankets and old lap robes. Jolt! Jolt! Jolt! Jolt! Every step of the way told them of the bitter cold, the frozen road, the strain on the tired horses; for most of the families I used to meet there was at least another hour yet to be faced.

College and the Experience of Nature

Before the days of automobiles, the surrounding countryside frequently played an indirect but important role in college education. Many of the colleges were located in small towns or villages; in getting about, the students were pretty much limited to their own two legs. I remember the Michigan Agricultural College of my boyhood, a thousand acres of beautiful campus, its large collection of trees identified and labelled by Professor Beal. It was set in the middle of rolling farm lands, flanked on the north by thousands of acres of marshlands just beginning to be drained, on the south by the winding Cedar River, of an almost perfect size for canoeing. The college's only effective connection with the outside world was an uncertain one-track electric trolley which carried one to Lansing, three miles away. On pleasant Sundays in spring and fall, the students swarmed over the landscape, through the beautiful old college woods, down the farm lane, up the Cedar River, out along the muddy corduroy road which crossed the Chandler's Marsh. When old graduates returned, decades later, it was frequently the Cedar River and the college woods they most wanted to see.

Just as the town, the river and the countryside played a real part in the education of Oxford and Cambridge men* so American college boys a half century ago were unconsciously learning from the villages and countrysides amidst which their colleges were placed. Williams had a "Hopper" day when Mt. Greylock was climbed by way of the Hopper trail, various other New England colleges had their "mountain" days and a "Dartmouth hymn" was written (and is still solemnly sung), which pays tribute to:

". . . the still North in their heart
The hillwinds in their veins"

(though friendship with numerous Dartmouth men leads me to doubt the anatomical location of New Hampshire granite ascribed to it by the song). Nor was this purely a New England phenomenon, though it had certainly been pointed up there by the determined cultivation of Nature which followed Thoreau's posthumous success. In early days at Berkeley, staff and students alike were acutely aware of their situation among bare grassy hills, facing the Golden Gate; for some decades the landscape of the Stanford ranch dominated the activities, and in more subtle ways the attitudes, of Stanford students.

The development of the automobile has so freed the American undergraduate from his spatial dependence upon alma mater as to confront college and university authorities with a complex set of disciplinary problems. Deans and professors have been occupied with such problems as keeping the student at the college and his automobile off the clogged campus roads. There has been little opportunity to think creatively as to how, with automobiles and good roads generally available, the wider landscape in which present day undergraduates move so freely might be used effectively in

*Charles Darwin's most effective education came from walks through fields and along roadsides with his botany professor. Lord Keynes, writing an official anniversary essay on Sir Isaac Newton, was careful to indicate, for other Cambridge men, precisely where Sir Isaac had most probably been lodged during his University career. In his satirical novel *Zuleika Dobson*, Max Beerbohm describes how at Oxford the river fog creeps into the hero's rooms at night, he goes on to contrast the ways of Englishmen conditioned through generations in universities set in low river valleys versus North Britons with their universities perched on rocky heights.

in education. Colleges situated in large cities have on the whole been quite as oblivious of the community around them and its landscape, as their sister institutions in smaller communities. Harvard long ago gave up Cambridge as a bad job and built walls and monumental gates to define more clearly its encysted status.

The Dartmouth Example

I can think of only one American college in which the landscape played a prominent role during the half century in which the automobile became more and more insistently a campus problem. Significantly, even in this example, Dartmouth, it was due to students and alumni though a number of faculty members were creatively sympathetic in the development of the Outing Club's activities. Though this program has its roots in the days before the automobile, the arrival of more flexible transportation extended rather than hampered Outing Club operations. Many a present day Dartmouth man knows, loves and understands the terrain of all central New England in much the way his grandfather was attached to the immediate countryside. Though almost entirely extracurricular, Outing Club activities (particularly in the Winter Sports program) have had a decisive effect upon many phases of Dartmouth life. They have, among other things, resulted in important Arctic research being centered at Dartmouth and have given new dimensions to Dartmouth teaching.

The Dartmouth solution, however, is, and must remain, unique. No other American college or university is so richly endowed with a long season of adequate snowfall, a terrain which provides lakes, rivers, and mountains immediately at hand. And there are other important factors. Few American colleges have had such a reputation for undergraduate vigor and enterprise. (Even before the days of its winter sports dominance, Dartmouth's football song was widely paraphrased to read "Dartmouth's in town again; run girls, run.")

It is a tantalizing problem to suggest how modern undergraduates might gain a wider and deeper education from the immediate landscape of their college days and the wider terrain brought close at hand by the automobile. Young people in general, certain young men in particular, like to explore; they have a strong internal drive towards this kind of activity, a drive which seems almost instinctive in origin. Good education works through the emotions quite as much as through the intellect. Many able young people who are scarcely touched by ordinary classroom experiences are readily awakened through their interest in the landscape around them. It is certainly highly significant that Charles Darwin, after a most undistinguished undergraduate career, was finally aroused by tramps through the countryside with his botany professor.

Exploring the College Landscape

The problem is tantalizing because there are so many different ways in which a solution might be attempted. The Dartmouth example suggests that indirect methods might in the end be more successful than merely going up to the student and saying, "Now here is some landscape. Go ahead and study it." It is tantalizing too because one has the feeling that a really good solution should be helped quite as much as hindered by the presence of automobiles. If, a century ago, undergraduates consciously and unconsciously derived certain values from the restricted landscapes immediately available to *them*, how much more might we now hope to achieve with hundreds of square miles encompassable within a few hours?

If the problem were handed over to me for a practical solution, I should be tempted to follow somewhat upon the path laid down by Sir Patrick Geddes in the studies of city and country which eventually led him and his students into town and city planning. I should like to try centering the work in a large classroom or hall with good light and as much wall space as possible. (An unused barn or barracks would be excellent.) With the class doing most of the work, integrated by conferences and regular lectures, we would get together for our own mutual education a kind of semi-temporary exhibition of the college's immediate landscapes and the hinterlands beyond. We would begin with the simple things immediately available, auto maps, city maps, state and national geological survey maps. We would explore the human ecology, the geology, the history of the area. We naturally might, since I am a botanist, look as closely as possible at one or two plants of particular significance for the area. In coastal California this could be the introduced eucalyptus and the weed oats; in Kansas, wheat and wild sunflowers; in New York City, the Tree of Heaven and one or two of the tropical aroids so ubiquitous in that city's living rooms. As the year went on, individual students and groups of students would be encouraged to go as deeply as possible into significant local situations, presenting their finding in reports and displays, with diagrams, photographs, and maps. Some displays might prove pertinent enough to leave from year to year. Others would be removed or revised before the end of the course. Wall space, large celotex panels, plenty of good thumb tacks, and a bus or station wagon would be our chief basic materials. Such a community exhibit could be the center for the work of an elementary class one semester and for an advanced seminar the next. Depending upon curricular needs it could be closely identified with biology or with geography or with history or with none of these or with all three together. Teaching it in such a way that dullards were awakened, that the brilliant and enthusiastic few did not do most of the work and get most of the benefit, that special skills (as in drawing or photography) were put to effective use, would be a stiff assignment for any teacher. Yet, if after some preliminary failures (or qualified successes), such a course could be taught effectively it would benefit the students, the college, the community, and the teacher himself.

The Cornbelt Farmer and the Cornbelt Landscape

As one travels north and west in the cornbelt, the landscape is more and more dominated by the windbreaks around nearly every farm home. One has only to glance at them casually to see how necessary they are for comfort; the trees and bushes themselves are (even to the summer visitor) eloquent of the struggle against the wind. On their northern and western edges the windbreaks have been shaped by it; branches and trunks bend away from it; the inner rows of trees are higher than the outer ones even though there may be a protecting outer screen of hardy close-twigged bushes, Russian olive, Tartarian honeysuckle, Osage orange.

Within this sheltering screen the farm-dweller of the cornbelt spends much of his life. What was to the pioneer farmer an open prairie is for his great-grandson a wall of brush and trees, a clutter of farm buildings (house, barn, chicken houses, machinery sheds, corn-crib) pig lot, a tree dotted lawn, a few flower beds, a vegetable garden, a glimpse of the road. In this sheltered nook he, his family, a good portion of his livestock, and his expensive mass-production machinery are pretty effectively cut off from the landscape of which they are a part. The children play amongst the alien weeds under the windbreak or in a sandbox in the yard amid the farm buildings. Neither for the farmer nor his family is there the constant identification with the sweeping landscape round about, which is enforced upon many a plainsman farther west.

Yet it was the prairie landscapes, with their copious wild flowers and tall waving grasses, rather than the short, drier grass plains to the westward, which so fired the enthusiasms of the first European visitors. Early naturalists, generally laconic in their accounts of what they did and what they saw, almost to a man wrote vivid accounts of the seas of waving grass, the tall blue-stem and the turkey-foot, when they first encountered them in what is now the cornbelt, the world's greatest mass production food factory.

Today the prairie grasses are gone, except in an occasional fence corner or along the railroad track, but the essentials of the old prairie landscape are still there to see, the gently sloping, subtly modulated surface of the land itself, the full dome of the sky overhead. For the last decade I have had a small experimental plot of corn in the middle of a much larger field at a corn-breeding establishment in central Iowa. By the very nature of my experiments, it has been necessary for me to spend a fortnight or so every late summer or fall, in examining corn plant after corn plant for various critical characters, recording the boiled-down technical facts in a field note book and harvesting the ears for next year's experiment. This takes a lot of time. Sometimes a full day's work moves us down the field no more than fifty or a hundred feet. I never begrudge the time; it is an experience I look forward to with greater joy each year. The strict order of the breeding plot has its own functional beauty, the straight rows of corn plants, each one numbered with its own number, the narrow "ranges," blocks of ten to twenty plants each, into which the field is divided by straight pathways at right angles to the rows. The ripening corn plants have their own symmetries. As the autumn advances and they get drier and drier, they fade into a curiously beautiful combination of colors. The plants themselves pass from a yellowing green to a yellow stalk with a graying leaf. The row upon row of Kraft paper bags (which protect the hand-pollinated ears from contamination) weather to a softer and grayer silver brown. The breezes and winds as they sift across the corn field take on a reedier and more sibilant tone as the plants become drier and stiffer day by day and week after week.

There is the gentle sweep and swell of the original prairie surface with these ordered plants stretching as far as the height of the plants and the dip of the land will let one see. There is, all day long, the full arch of the sky. It is seldom completely empty of clouds; overhead it may be clear, but off to the north or the east will be a bank of clouds or an advancing or retreating strip of lozenge-patterned mackerel sky. Even with little technical knowledge one becomes aware of the swirling air-masses around our planet, the dramatic advances of cold fronts, the up currents during hot sunny weather with their puffy little fair-weather clouds which grow during the day from the steaming heat of the prairie itself and are gone by sunset.

Few indeed are the moments when the modern prairie farmer is alone with these immensities; in the daily run of things he may go weeks or months without such an experience. During the night, at mealtimes, and for a considerable portion of his daily chores he is within the sheltering tangle of the windbreaks. When he goes out onto the farm it is in company with expensive, high speed, noisy, mass production machinery. He is operating (or helping to operate) tractors, cultivators, spray equipment, field choppers, hay balers, ditch diggers, combines, seeders, corn pickers, bulldozers, manure spreaders, seed drills, dusters, post-hole diggers, hay loaders. These are the intricate machines which allow him and his family to produce more useful molecules of foodstuff per acre, per man hour, than has yet been achieved elsewhere. These are clever modern machines; they demand attention and care. The cultivators, for instance, are operated faster than a man can walk, yet the farmer's son riding aloft under his big umbrella, drives them so accurately that the cutting blade buries weeds right up to the base of the stalk but does not come close enough to injure the corn plant itself. Or watch the driver of a one-man hay baler as he picks up the dry grass, rolls it into a neat bundle, secures it firmly, and then disgorges it onto the short meadow, steadily moving down the field all the while. The operator of such machinery is too busy looking down to spare much time to look up. He has only a little more chance for acquiring a sense of the landscape around him than any skilled operator of complicated machinery in any other modern mass-production factory.

Confederate Violets

It was from the Confederate violet that I got the most dramatic demonstration of how one may unconsciously encourage those camp-follower plants which travel about the world with him. The Confederate violet is a handsome blue and white variety whose exact history and range I wish I knew more about. I have seen it growing with little or no encouragement (sometimes even in spite of being fought as a weed) in numerous semishady gardens of the Middle West and Upper South. Characteristically it belongs with those old white houses with spreading porches and shaded lawns. At such homes you not only find it here and there in the garden and among the shrubbery, but it tends to carpet the more or less untended areas between the vegetable garden and the back of the lot. It is large-flowered, for a violet, with a deep blue center which fades out so gradually into the off-white petals that the casual impression is of a blue and gray flower. It is fairly widely planted even outside the South; I remember having seen it in the front yards of several old houses opposite the campus of Amherst College. It was growing in under shrubbery and here and there at points where the grass was thin, with the general air of something which had spread in by its own efforts and was then much too charming to be rooted out completely.

In St. Louis it is a little too much at home; it can almost be a pest in shady places. For that reason I have never planted it in my town garden, but we have a woodland camp which had a single large plant of it, set out by the previous occupants, and there I learned about its likes and dislikes. Our camp is 'way off from everywhere, in what was originally a barn. There is a small cleared area that is blue every spring with the native violet and in the middle of this open space we do our cooking during nice weather. We wash our dishes out there in the open air and there are natural hollows, sloping away into the woods, where we throw out our dish water. One of these became more shaded each year as the trees spread their branches wider and wider overhead. As the shade increased there was less blue grass each year and more violets. There were thousands of native blue ones around the edges of this spot and only a single plant of the Confederate variety up against the side of the barn, yet the violet carpet is now composed quite as much of the Confederates as of the native sorts. They have spread out amazingly, being apparently one of those violets which reproduce their own kind exactly from inconspicuous little seed-poddy blossoms which come on after the showy ones have gone by. They do not occur evenly all over the place but are concentrated right where sudsy dish water is most likely to have been thrown. Nor do they remain the same from year to year. If we use the camp a lot there are more Confederate violets the next spring; when, as during the war, we are away altogether, the native species begin to get the upper hand. In one year it was almost like a laboratory experiment; a little triangle of blue and white, fitted exactly to the area of dish water runoff in the midst of a carpet of deep blue.

Were I a much younger man I would like to make these plants the subject of some fairly precise experiments, tracing down the connection between the dish water and the violets. Do the Confederate violets benefit directly from something in the soapy water? I would suspect not. The whole picture of where they do and do not grow without encouragement suggests that they grow in such places not because they really prefer dish water, but rather because they can put up with it and the native violets cannot. Even more intriguing would be an attempt to narrow down the exact relations between the dish water and the violets. Is the effect directly upon the violets or indirectly through various microorganisms in the soil? The latter hypothesis would be worth looking into. Any soil is a set of communities as well as a set of substances. It has bacteria, protozoa, algae, and other microorganisms, the balance between them greatly affecting its suitability for various sorts of plants. Traces of greasy substances have a tremendous and differential impact upon these several kinds of microorganisms. The soils about towns and villages frequently have a puddled look, as if the algae in the soil had somehow got the upper hand and increased out of all proportion to the other members of the soil community. It is in precisely such soils that the Confederate violet seems to be at an advantage. A few well-planned experiments, carried on over a decade, should yield data for pinning down this fundamental problem.

Where did the Confederate violet come from in the first place? I wish I knew. It probably turned up in some garden and was kept because of its unusual color. Its ability to tolerate certain slightly corrupted soils has insured its survival and aided in its spread. It is similar to one of our commonest eastern American violets, yet not quite typical of that species. If anyone has found the original wild strain from which the Confederate violet spread into the dooryards and garden edges of the mid-South I have not heard of it, in spite of asking many questions and keeping my eye out for scientific articles dealing with such matters.

Where and how are such plants generally bred? There are as yet few data which bear directly on this problem. Most of these plants have been with us a long time and, while new ones still turn up, one cannot hope just any day to find a plant which is, right now, in the process of becoming a camp follower. The clearest and best documented case which I know concerns two blue salvias from coastal California. They were first studied by Carl Epling and his students. This example seemed such a critical one to me that I too have studied the salvias intensively, choosing them as demonstration material for my classes in which we were analyzing the ways in which evolution proceeds at the present moment.

These two blue sages grow more or less intermingled over wide areas in southern California. Though they are both members of the genus Salvia, they have different-looking flowers and the plants themselves are strikingly unalike in general aspect. *Salvia apiana* raises its flowers aloft in dramatic wands of bloom which shoot up into the air three to seven feet and more. The individual flowers are conspicuous, the expanded lower portion of each blossom serving as a retractable landing platform for visiting bumblebees. *Salvia mellifera* is a much more ordinary-looking plant though it has an air of refinement which is lacking in the lustier *apiana*. Its tiny, trumpet-shaped flowers are a bluer blue, being not very much larger than the cone of exposed black lead on a well-sharpened pencil. They are borne in tight circular bunches which are stacked, pagoda-like, one above another on low, much-branched bushes. These two species grow, frequently intermingled, on sunny slopes, rocky or gravelly for the most part, along roadside banks, on the steep southern faces of the coastal mountains, particularly along mountain roadways, often persisting here and there in various remnants of native vegetation well within metropolitan Los Angeles.

Though they are conspicuously different in size and shape, these two salvias hybridize; an occasional plant of this distinctive, vigorous intermediate can usually be found by any naturalist who makes a determined search. Epling learned that in the experimental plot such hybrids were easily produced by cross-pollination. Furthermore, these artificially produced hybrids were fertile, and under experimental control would readily yield a varying lot of secondary hybrids, combining in various ways the characteristics of the two parents. However, such hybrids are seldom found in nature and the few which were observed were largely limited to sites where man had greatly altered the face of the landscape.

It was at such a point, at the very base of the mountain wall, that I made an intensive analysis of the relationships between *apiana* and *mellifera*. A Forest Service pathway wound more or less horizontally over the rugged, sunny slopes and along it these two salvias were frequent. Where the rocky flanks of the mountains jutted out most abruptly and the soil was almost pure gravel, *Salvia apiana* was the commoner. Along the very keel of these ridges it was often the only plant of any size. In between these outfacing ridges, *Salvia mellifera* was more abundant and sometimes made dense little thickets several square yards in extent. For a mile or so of pathway there was no outward evidence of hybridization. There were many *apianas* and *melliferas*, but to all except the most exhaustive examination they seemed to be perfectly normal *apianas* and perfectly normal *melliferas*, varying a bit from plant to plant but showing no influence of any mongrelizing. However, an exhaustive study of the plants immediately along the path modified this snap judgment very slightly. The plant-to-plant variation between the *apianas* showed no relation to *Salvia mellifera*, but among the *melliferas* themselves virtually all the measurable plant-to-plant differences could be accounted for as the greatly diluted effect of previous hybridization with *apiana*. The *mellifera* plants which were most unlike the norm for that species were precisely the kinds of plants which would be expectd if about one ancestor out of eight were from the other species—in other words, if a hybrid with *apiana* had crossed back to *mellifera* and then these three-quarter bloods had again been back-crossed to the same species. The differences from the norm among the *melliferas* were *all in the direction of apiana*, and those which, for a *mellifera*, had somewhat *apiana*-like pubescence tended to have, as well, flowers and inflorescences more *apiana*-like than is normal for a *mellifera*. These very slightly mongrelized *melliferas* seemed to be more abundant at those points where campers had most abused the native vegetation.

At one point the path swung down toward the base of the mountains. Here a very different relationship between the species was to be seen. Originally live oaks had abutted on the sage-covered slopes. Long ago most of the oaks had been cut down and in a small rectangular area olive trees had been set out at intervals. This little orchard had then been abandoned and native vegetation had spread in among the olives. There were many sages, and no two of them alike. There were a few that were suspiciously similar to the hybrids raised by Epling; there were many more that resembled the hybrid descendants which he had grown in his experimental garden. There were, in particular, a very large proportion of plants closely similar, on the whole, to *Salvia apiana* yet showing obvious resemblances to *mellifera*. This is particularly to be noted since in Dr. Epling's experience such plants are seldom met with in nature. Most of the mongrels which he and his students found growing wild were, like those I found along the pathway, *Salvia mellifera* with a barely detectable trace of *Salvia apiana* in their ancestry.

Well, so much for a review of the essential technical facts. Two species which characteristically grow more or less intermingled over wide areas seldom produce any hybrid descendants, even though they are easily hybridized in an experimental plot. However, when a strange new sort of habitat was made available close at hand, hybrids and mongrel descendants filled this area to the virtual exclusion of pure forms of either species. The point to be emphasized is that the mongrels not only were virtually limited to the abandoned olive orchard, but in that circumscribed area they were not merely common—they practically made up the entire population of sages.

All of this indicates (and the evidence is particularly good because so many of the features have been independently verified by Epling and his students) that hybrids and various grades of mongrels must have been constantly produced on the mountainside but that very little of the seed came up, or if it did germinate, it was quickly eliminated. The coastal sages and their associates are a well-established evolutionary unit. For centuries, probably for millenia, they have been selected and reselected to live effectively with one another and with a characteristic set of associates. Nearly anything different (like an intermediate hybrid, or a seven-eighths mongrel) is no improvement and very few of these variants ever succeed.

However, when among the abutting live oaks the trees were cut down, a new area was opened up to the sun-loving sages. Olives were planted at well-spaced intervals and then the whole tract was abandoned. Here was a whole little plot quite unlike anything previously known to these organisms. It had, like the mountainside, a good deal of sun, but the soil, in large part, was like the soil of the live-oak woodland. The olives were something new and they were widely spaced. Some of the plants which first spread in must have come from the live-oak woods, some from the mountainside; others were probably weeds. There were many new kinds of places for plants to fit into and these plants which were coming up together were not just the usual lot of sages and their associates or oaks and their associates. Here was a spot at which something new and different might be at a selective advantage. When two species are crossed, the number of different new kinds of combinations which may arise from their hybrids is staggeringly large. (We see a little something of what it might be like in the thousands of named varieties of bearded irises which have arisen in the last few decades from crossing a very few species of wild irises.) For the various new niches which arose in the abandoned olive orchard, thousands of kinds of hybrids were available as soon as a single first-generation hybrid began to shed its seeds over the plot. Most of these were no more suited to the orchard than they had been suited to the mountainside, but a very few of them did fit in. From them a new crop (or perhaps successive new crops) of even fitter mongrels could be selected. Had there been enough abandoned olive orchards adjacent to enough mountainsides, it is conceivable that a dozen or more new salvias of mongrel ancestry might eventually have been selected out of the ferment.

Since man came on the evolutionary scene he has been increasingly a maker of new kinds of habitats. The abandoned olive orchard which replaced the live-oak forest below the sunny mountain slope is a typical example of his day-to-day activities. It has certainly not always happened that he made these strange new places immediately adjacent to an area where previously unwanted hybrids were being more or less continuously spawned. It seems likely, however, that out of the thousands of millions of times when he created such disturbances there were a significant proportion when he did indeed create new opportunities for previously unsuccessful hybrids and other variants. In so doing he all unconsciously bred his own weeds and camp followers. In some such way the Confederate violet probably had its origin.

Maize of the Southwest

Maize of the Southwest; it has been there for over two millenia. It has felt the impact not only of the various peoples that flourished there but tardily and dilutedly of great civilizations whose type of corn traveled far outside their empire boundaries. It has been grown there by very different peoples, by Basket Makers, Hohokam, Pueblo, Papago, Spanish landowners, Navajo, Yankee pioneers, USDA cornbreeders, interior decorators. Let us begin by discussing the interior decorators. At the moment they are having the greatest immediate effect upon the complexion of southwestern maize.

It is easy to smile at them and at the new direction they have brought to corn breeding in the Southwest, the brilliant ears with varicolored kernels, the braided strings of gaudy maize which travel back in tourists' autos to grace living rooms and patios from Cape Cod to San Diego, the even more staggering amounts that move here and there over the nation to antique shops, to purveyors of supplies for decorators, to florists; eventually to emerge in the full light of publicity at Kiwanis banquets, Thanksgiving cocktail parties, and in decorative arrangements at National Flower Shows. It is easy to smile, knowing that the bulk of Indian maize in the Southwest was plain colored up until the last decade and that many of the distinctive Indian kinds were as carefully selected and as rigidly true to type as any Corn Belt variety in the days of the champion ear and the champion bushel.

It would be easy to smile but it would be wiser to recognize this fad for maize-to-look-at as a sign that Americans are at last becoming real natives of the continent which they have conquered. Maize is not only fantastically productive, it can in sympathetic hands become fantastically beautiful. Pre-Columbian maize in the areas of highest culture was certainly bred as much for its looks as for its yield. It was the center of religious life, it was venerated, it was one of the avenues for invoking the presence of God. It would, I suspect, have been blasphemous to refer to it then merely as the "staff of life." If, in even trivial ways, we are beginning to enjoy corn just for looks, we are to that extent becoming worthy trustees of the civilizations from whom we took over this fantastically variable giant grass which we Americans call corn.

So variable is maize that more than any other crop plant it reflects the histories and the attitudes of the people who grow it. The history of maize in the Southwest, when completed, will be a history of southwestern civilizations. Technical papers on southwestern maize have come out with accelerating frequency in the last two decades. The data roll in. Old controversies are settled; new controversies arrive. Heated and sarcastic phrases find their enlivening way into cool scientific prose. Out of this yeasty ferment will eventually come a pure and heady distillate of truth. The time for that refreshing draught is still well in the future, but a few generalizations are becoming increasingly apparent. Maize in the Southwest has been affected by various peoples, at various times and from various directions. One of the most amazing of the fairly well established details is that during the first millenium of the Christian era, Mexican influence filtered slowly into the Four Corners Area decade after decade and century after century, *and from the North not the South!*

Corn Before Columbus

The corn of the Southwest of day before yesterday, the day before the decorators, was simple in its overall relationships.

Raiding tribes, with their catch-all harvests of this and that blurred the picture. The Navajo in particular with their love of color and their adaptability had taken corn from various Pueblos and from various traders and had produced useful blends, some of them delightfully colored. Old Spanish-American communities such as the one at Chimayó had brought up varieties from Mexico and Central America to breed them into new sorts which were as colorful as they were productive. With these exceptions the story was a simple one and had remained with very slight changes since late pre-Columbian times. The eastern Pueblos had maize in many ways more like that of the eastern Indians, long eared, big at the butt end, with wide, crescent-shaped kernels. The farther west one went among the Pueblos, the less eastern influence could be detected, until among the Hopi and the Zuni one found various specialized varieties well suited to the environment, varieties which had persisted there virtually unchanged for almost a thousand years. It is today difficult to give an up-to-the-minute account of Pueblo maize. Modern autos and modern highways have broken down the isolation of the Indian communities. Since World War II the maize of the Pueblo Indians has apparently changed as much as in the previous half century. This is true even of such conservative Indians as the Hopi whose cultural continuity with their pre-Columbian pattern of living had never been broken. Up until World War II any self-respecting Hopi farmer grew at least three quite different kinds of maize and some grew as many as seven. The commonest sort was a white flour corn; large squarish white kernels, so soft they were almost like little plastic bags of white corn-starch, set on a short flattish ear. This was the backbone of the diet; it was used for green corn; when ripe it was made into a fine meal or flour and cooked in various ways. The second commonest was a hominy corn; long narrow ears usually with flinty kernels of a steely gray blue. One of the most fascinating types was the dye corn grown for its coloring matter as well as for food. The plant was in many ways more like the primitive types grown by the Basket Makers. The kernels were small, ear and cob alike of a purple so deep it was almost black. One or two of the kernels boiled up in a cup of hot water give it a beautiful wine-purple tint. The dye corns were used, among other things, to color the famous *Piki* bread of the Hopis, one of the tastiest and certainly the most beautiful of New World foods. Bread is a misnomer; even wafer would be an understatement.

Hopi Corn and Its Variations

Today fewer Hopi grow the old varieties, though *piki* bread is still made and dye corn is still grown. Food colors for the *piki* bread are bought at chain grocery stores and some Hopi raise bright colored corn obtained from the Navajo, to sell to tourists. Conditions are changing so rapidly that one would have to be in almost continuous touch with the Pueblos to discuss the practices of the current year authoritatively. The conservative Hopi were amused or contemptuous at their first exposure to the American demand for the parti-colored ears desired by tourists and decorators. They always rogued their seed stocks drastically. The white corn should be white, the blue blue; only when a variety was typified by variegation or by a mixture of white and blue kernels on the ear, were piebald varieties deliberately encouraged. So careful were the Hopi that traditionally at harvest time any ears with even a single off type kernel were laid at one side to be used for food. Hopi farmers were amused to see these second-rate ears with off-color kernels picked out by tourist visitors as most typically Indian. However as the demand increased and as the prices went up it was easy to supply this strange new market. (Some of the brightest Navajo blends were certainly produced by allowing the Hopi dye corn and the Hopi hominy corn to interbreed.) By planting different varieties close together one increases the chances for cross-pollination and the production of off-type kernels. By planting the off-type kernels the following year one can obtain hybrid ears with a great variety of motley colorings. It is too bad that someone with a technical knowledge of corn genetics has not had the time and the opportunity to make a survey of the strange little business of supplying souvenir corn to tourists which had been developing since World War II.

The Return of the Oldest Corn

Interesting as is the maize of the Pueblos it is only part of the story of corn in the Southwest. Far from the transcontinental highways and railroads and therefore known to comparatively few tourists, the Papago and their neighbors carry on one of the world's most remarkable agricultural civilizations. In the desert south of Tucson along the International boundary these clever agriculturists grow ancient crops, specialized kinds of corn, beans, and squashes which will produce a usable harvest on fewer inches of rainfall than are used anywhere else in the world. Papago corn would be scorned by most decorators, though it has its own peculiar beauty. It is mostly all the same color, a pale straw yellow. Archaeologically it is of supreme interest because it differs little if at all from much of the maize grown by the prehistoric Basket Makers. It has the same great variation from ear to ear, the same irregular kernels tightly held to a narrow, bony cob. From the culinary point of view it is superb. It has a soft floury texture and produces a light yellow meal which makes superlative spoonbread. Unlike many of the Pueblo varieties it is not hard to raise in the North. Having been planted for centuries not at some regular date in the spring but whenever the first good summer rains came, it has had bred into it a remarkable independence of day lengths and seasons of planting. It bears as well in my garden in Missouri as it does in Arizona and even gives a good account of itself in central Iowa where my experimental plots are located. For over a decade we have been selecting it for greater productivity. It is not yet a real commercial success but it makes a corn meal so tasty and so light that my miller grinds it for nothing, just to get some himself and a little to pass around to one or two old friends.

Sunflowers in the West

From the Dakotas to New Mexico, from eastern Wyoming to western Iowa, few plants are so characteristic of the Great Plains as the common sunflower. From mid-June to mid-September on stems sometimes scarcely a foot high, sometimes six feet and more, they line the highways and the railroad tracks, they grow along the irrigation ditches, they come up in vacant lots, they fill sandy blow outs with resinous leaves and golden flowers almost to the exclusion of all other weeds. They are a mixed lot, but it is a relatively simple mixing of two main elements, one early, low, small-leaved and addicted to sandy places; the other late, tall, large-leaved, demanding richer soil.

Early Petiolaris, Later Annuus

There seems to be no generally used common names for these two extremes, more's the pity, and if we are to discuss them with any precision we shall have to use botanical Latin, *Helianthus petiolaris* is the small early-flowering species; *Helianthus annuus* its larger, later-flowering relative. If we drive across the Plains in June, we shall see little but *petiolaris* in bloom. By mid-August there will be as much *annuus* as *petiolaris* in many places and when the season ends *annuus* will be in the lead everywhere.

Characteristic plants of the two species are easy enough to tell apart. It is the three-quarter bloods and their descendants, the plants which have one ancestor out of eight or one out of sixteen which belong to the other species, that make exact descriptions difficult. Stop your car almost anywhere in the Great Plains and gather one flower each from fifty to a hundred different plants and you will probably pick up a few. Here in a sandy spot where the road scraper dug up soil for grading the highway are *petiolaris*, about two feet high, with narrow leaves. Further down towards the railroad track in richer soil are taller coarser plants just now coming into flower. These are *annuus*, or so close to it that if we want to give them any name it will have to be that.

The Hybrid Species

Petiolaris is smaller, it is almost daisy-like, the ray flowers are a paler yellow and the central disk is very small indeed. A sunflower is not a single flower; it is a whole head of miniscule florets and back of each one is a stiff pointed little straw known as the chaff. When the head begins to flower, the points of the chaff are all slicked down flat, pointing towards the middle; as flowering proceeds one circle after another straightens up and the tiny florets emerge and open. The chaff is dark red-purple so dark we may call it black for all practical purposes. In *petiolaris* these tiny bracts bear conspicuous white hairs at the tip; in the center of the flower head where all the tips come together this produces a lacy gray-white spot. *Annuus* coming into flower is therefore a big deep yellow daisy with a black center, *petiolaris* a smaller, lighter yellow daisy also black-centered but with a white eye in the middle of the black. If we bother to study them plant by plant, we shall probably find that at least a half of the *petiolaris* are more or less alike but that the rest of them tend to be larger, and deeper yellow with bigger disks and with less conspicuously lacy centers. Similarly we shall also find off-type plants among the *annuus* collections. They will have smaller flower heads than good *annuus* with smaller central disks; sometimes we may find one plant or two with definite indications of lacy white hairs towards the center. Nearly always such an exception will give indications in various ways of its partially mongrel ancestry. The rays will be narrower and lighter and fewer than are characteristic for *annuus*. From the most *petiolaris*-like of these plants it will not be much of a jump to the most *annuus*-like of the off-type *petiolaris*.

From hundreds of such observations, as well as from Dr. Charles Heiser's careful experimental studies, we know that the Plains' sunflowers are a double-barrelled complex. The two species hybridize rarely but from these hybrids crossed back to the parents and then crossed back again and again, comes most (if not all) of the perplexing variation.

Stop sometime for an hour or so and study a whole roadside swarm of these golden weeds; your senses will be sharpened. The sea of yellow flowers which used to stretch out endlessly and uniformly is now differentiated. Sunflower heads will have acquired individuality, a sunflower population have become as absorbingly varied as a human population. You will also have made the first step towards acquainting yourself with one of the keys to the history of the high plains. For these sunflowers are of more than passing interest. It was out of this rich ferment of intergrading species that the cultivated sunflower was born, certainly in pre-Columbian times, presumably before maize had reached the Plains. It was this same ferment which bred weediness into the Plains' sunflowers. The yellow lines of sunflowers bordering the highways are obviously pretty recent. When did they first come in along trails and about encampments? Where were they and what were they like when early man first reached the Great Plains? These are the questions which a few of us who study sunflowers are beginning to ask ourselves.

The archaeological evidence is clear. There are seeds of both wild and cultivated sunflowers from various prehistoric sites in the Great Plains and the Southwest; very occasionally there is a whole dried flower head. Sometimes there are sunflower images. Kidder and Guernsey uncovered sunflower decorations for a whole room. They are life-size models of wild sunflowers, fashioned out of wood and out of leather and though now somewhat faded they still show realistically the yellow rays and dark central disks.

There is a good deal of scientific spade-work to get out of the way before the archaeological evidence can be utilized to the full. Only very occasionally will the sunflower remains from an excavation be adequate enough to determine from casual inspection just what kind of sunflower was being grown. Sometimes there may be no more than a single carbonized seed. Such remains, however, have microscopic details which are significant, or rather they will be significant when Dr. Heiser and his students have made wide enough surveys of cell types in all kinds of sunflowers.

Native of Southwest or Plains?

Yet a good deal about the history of the sunflower can be determined just from a study of those now growing on the Plains. It seems fairly probable, for instance, that *petiolaris* itself reached the Plains ahead of man. Though it is commoner where man has been disturbing the landscape it comes far closer to having the look of being native than does *annuus*. The latter is almost certainly post-human on the Plains. It is one of those plants which travel around with man, which he had unconsciously bred to fit into the scars he makes in the landscape. What was its origin? Before too long we should have an answer. If it is truly post-human, then the species it was bred from must still be with us. If it is pre-human or partly pre-human then we should find somewhere in the Plains or the Southwest, a landscape in which it and its hybrids are not confined to man-distributed habitats. Near Shiprock, New Mexico, I once studied a population of sunflowers which were growing well away from the road in a low basin where the water had stood after a rain. Perhaps pre-human *annuus* came from such a spot. In the almost succulent texture of its dark green leaves *annuus* suggests a plant native to salty or alkaline areas. Perhaps before it first began to travel with man and get mongrel vigor and mongrel variability into its system it was restricted to such little *playas*. Perhaps.

Anatolian Mystery

Though when I went to the Balkans in 1934 I knew little about such cucurbits, I was fascinated by the squashes and the pumpkins I saw in Bulgaria, Rumania, and Yugoslavia. They were brought in from the fields by hand and by the cartload. They made great piles in the farmyards, they were on sale in every little market, and everywhere I saw them they varied enormously. They were green, yellow, orange, white; striped, blotched, marbled, speckled, or uniform colored. They were quite round, or much wider than long, or several times as long as wide. In the two decades since this Balkan trip I have visited many native markets in the New World. There have been squashes and pumpkins in most of them, seldom in such abundance, never in such bewildering variety.

  Nor is this just one visitor's casual impression. Vavilov, the Russian student of cultivated plants, after leading expeditions to the back corners of Asia, Africa, and the New World, found that the variability of the world's cultivated plants is concentrated in a few ancient centers of which Anatolian Turkey is one of the most important. The variability of the squashes and pumpkins which I noted in the Balkans is even greater east of the Bosphorus in Anatolia. Though these cucurbits come from the New World and were found here in great abundance by the earliest European explorers, Vavilov's precise methods established Anatolia as the chief center of diversity for one of the species along with various crop plants of the Old World.

A quarter of a century later, Dr. Jack Harlan's Turkish collections indicated that other American domesticates are also centered in Anatolia in great variety. Dr. William Brown and I, working with Harlan's collections, have monographed the Turkish popcorns in great detail; the sunflowers are being studied by Dr. Charles Heiser.

When and by what routes did American food plants get into Turkey in such variety? For popcorn and tomatoes my students and I have begun to look into the evidence. These studies raise more questions than they answer, but the data, taken as a whole, are consistent, and they do point up the problem.

Tomatoes from the Mediterranean

Let us consider the tomato story first; it is the simpler of the two, though fragmentary. It began with an attempt by my student George McCue to work out the history of our everyday uses of the common tomato. A century ago my grandmother in her childhood knew tomatoes as "Love Apples," a supposedly poisonous plant grown as an ornament and a curiosity in the flower garden. In my own childhood the common country uses of tomatoes were pretty casual. They were used as a salad in the summer time and as a stewed vegetable at infrequent intervals during the winter. Tomato catsup was not made in our home, though tomato jam and chili sauce gradually made their entrance. I was in high school before I even heard of anyone drinking tomato juice, of college age before I tried it myself. It was not until I grew up and made Italian friends that I became exposed to what is now becoming the common American use-pattern, tomatoes in some form or other nearly every day throughout the year; tomato catsup bottles on the restaurant tables; tomato juice for breakfast; the mass production (in California and elsewhere) of Italian-type tomatoes, meaty, medium sized, adapted to canning, drying, preserving in oil so that they can be used throughout the year.

By a great variety of evidence McCue was able to show

that we Americans got our knowledge of the tomato's proper place in the diet from our French and Italian immigrants, and that their everyday use spread from such centers as Charleston and New Orleans. His investigations reached the edges of far larger problems. When and where did the Italians learn about tomatoes? In the early 1500s apparently, and from the Turks. Just as the Italian common name for maize is *granoturco*, "grain of the Turks," so the tomato was originally called in Italian the "apple of the Moors," *pomo di moro*, a name which was shortly corrupted into *pomo d'oro*, "apple of gold."

It is to our Italian immigrants rather than to Mexico that we owe the solid flesh in our commonest American tomato varieties, even though Mexico is their ancient home. The very word tomato is derived from the ancient Aztec *tomatl*, but few of the Mexican varieties have solid meat. They are watery fleshed, they do not keep well; I have never seen or heard of their being dried down into a paste for winter use, except by Italian-Mexican familes. For this pattern of use, for this kind of tomato in great variety, one must go to the Mediterranean. From Italy across the Balkans and all the way into eastern and northern Persia, in many of the islands of the Mediterranean the tomato is used both as a fruit and a vegetable. There are special meaty-fleshed varieties which are dried in the sun on flat roof-tops or pickled for winter use in salty brine, or boiled down into a paste and stored in pottery jars with olive oil across the top to keep out the air. It was from the Italians that New England plant breeders got solid-fleshed varieties which they used in the early 19th Century for breeding improved kinds for the American market. A century later it was with these same kinds of Mediterranean tomatoes that California growers moved the center of the American tomato-canning industry to the West Coast.

How did the tomato, along with chili peppers, tobacco, big-headed sunflowers, squashes, (and, of all things, popcorn) get so thoroughly established in the eastern end of the Mediterranean? When and where were these techniques worked out for keeping tomatoes through the winter; when and where were the solid-fleshed sorts first developed? If one had a detailed map of tomato types and uses for Turkey we would at least be able to phrase the question more precisely and perhaps know where to look for the rest of the answer. One thing I *do* know: this tomato center does not extend to India. Though its use there is increasing, the tomato is obviously just beginning to make its way into the Hindu diet. In those parts of India which I have visited it is much commoner in city markets than in village markets.

Popcorn in Turkey

For popcorn, however, I have been busy making just such a map; it raises questions so disturbing that some scholars prefer not to face the evidence. I made this map mostly from interviews with immigrant families and with visiting scholars, since popcorn is such a seemingly unimportant item that it scarcely ever gets into a travel book.

The map of those parts of the world where popcorn is as common in the villages as in the city and has been for at least some hundreds of years, begins in northwestern Mexico. It runs down through Guatemala and Honduras where through large areas it is so much an Indian trait (rather than a Latin one) that many educated people know nothing about its prevalence in their own country. It practically disappears as northern Panama is approached, to turn up strongly again among the savage tribes of Darien. Through Colombia, Ecuador, Peru, Chile, its use is common and ancient and well documented.

Halfway around the world popcorn turns up again with the same pattern of being a back-country habit, either unknown to city sophisticates or known only as a modern introduction in movie palaces. In southeastern Asia it is the aborigines back in the mountains who are the popcorn raisers and users, in the mountains of Indo-China, in Siam, in Burma, in Nepal, Kashmir, and on into Persia. In Teheran an observant American family can live for several years and never hear of it, but native families know all about it and send their servants down to the poor part of the city to buy it for the children. In smaller provincial cities on the Caspian Sea or in villages among the mountains on the Turkish border everybody knows about it, even the casual visitor.

Across Turkey, popcorn is in seemingly every village and town in great variety of ear types and kernel types and habit of plant growth. It can be scarcely knee high or much higher than your head. It continues on across the Bosphorus and goes up into the Balkans as far as the Turks penetrated and no farther. Around Sarajevo it is grown by both Mohammedan and Christian farmers, yet in southern Austria and northern

Yugoslavia it is virtually unknown. Along this whole east-west belt it has a variety of names, mostly different in every country, each name with a natural ethnic distribution.

Even more disturbing to commonly accepted theories is the map of those areas where popcorn *is not grown*: areas where people do not know how to use it even when it comes into their hands. It is sometimes grown, for instance, in Spanish villages, but merely for chicken feed. Those who grow it do not know how to pop it or realize that it *can* be popped. The special uses of popcorn are quite as unknown in Portugal or Spain as in Germany and England. If, as has been commonly supposed, the Portuguese and Spanish carried it from the New World to the Old, how could they have passed on these specialized varieties and a knowledge of their specialized uses without themselves having learned anything about it? For the Spanish with their aristocratic disdain of the dirty details of vegetable growing, this is a conceivable but scarcely credible hypothesis. For the Portuguese with their highly specialized horticulture, their enthusiasm for unusual vegetables in fine variety, their special flair for agricultural botany, this approaches the impossible.

For squashes and pumpkins, common as they are, we know so little about the world distribution of their species and varieties that we cannot suggest the route or routes by which they reached Turkey. For popcorn we have in the last two decades learned so much about the races of maize (which has an overall variation pattern not so different from that of mankind) that we can begin considering the problem. From Indo-China to Siam to Burma to Nepal to Kashmir the popcorns grown by aborigines show no close relationship to the kinds of maize taken around the world by the Portuguese and Spanish. They are, instead, very similar to some of the ancient popcorns grown in South America in prehistoric times and now become rare there, except in a few remote localities. The popcorns of northeastern Turkey are of this sort. A variety collected by Dr. Harlan in Samsun on the Black Sea, differs no more from a collection of this variety made in a Kashmir village, than one strain of Golden Bantam sweet corn differs from another.

However, the popcorns of the Aegean coast of Turkey are not of this sort at all, but resemble certain primitive varieties from the Caribbean. In other words, it looks as though the popcorns of Turkey owe their great diversity to the mingling of two highly diverse strains, one which came overland from the Orient, another which arrived by the sea.

Who Brought Corn to the Old World?

When and how these popcorns came to Turkey is a problem we can scarcely even discuss until all kinds of evidence have been thoroughly sifted. Few, indeed, are those western scholars who read Arabic fluently. Those who, like Jeffreys, have dipped into this literature have been amazed at the evidence they have found. He believes that the Arabs had maize in their possession by 1200 A.D. and that they carried it with them down into Central Africa, long before the Portuguese brought it to coastal areas. He is beginning to back up his hypothesis with linguistic maps. The whole problem is bedevilled by the close similarity of maize and sorghum (which is certainly native to Africa) and by our vast ignorance of the classification and the history of the sorghums of the world.

That Anatolia is the center of diversity for squashes and pumpkins and for popcorns is a good, hard fact. It has been documented by Vavilov for the cucurbits, by Brown and myself working with Harlan's magnificent collection of Anatolian popcorns. There are preliminary indications that sunflowers and possibly other American crops will show similar diversity there. The proper explanation of this fact remains a mystery and a challenge. It is a challenge which can be met by more meticulous study of the cucurbits and of their uses, by a rigorous sifting of the Arabic literature, particularly the geographies and the herbals, by linguistic evidence, by setting out to analyze the variation patterns of these plants down to the last piddling detail.

The City is a Garden

Perhaps if I had not lived comfortably in the very center of a busy Mexican town for six happy months, I should not be so conscious of the psychotic way in which we Americans do not accept the cities we have built. With each successive decade we spread farther and farther into the outer suburbs, trying to gain the kind of open country and grassy lawns that our grandfathers knew, leaving behind us wider and wider zones of blight and decay. The average American city now resembles all too closely a spot of blue mould on an orange. There is an outer zone of clean, bright new growth, advancing rapidly. Inwards from these outer fringes the growth gets denser and darker until in the center there is rot and decay where the actual substance of the orange has been eaten away, the mould is rapidly dying out, and various other microorganisms are crowding in to complete the dissolution.

For this ominous condition, the amateur Thoreaus and the professional naturalists of our culture are more fundamentally to blame than they realize. They have in the United States raised the appreciation of nature to a mass phenomenon, almost to a mass religion; yet at the same time they have refused to accept man as a part of nature. For the beneficial contemplation of the world around us, they would have us always get as far away from man as possible. They go to seaside and mountain top, at the very least to a farm, there to gather the values that our immediate forefathers somehow distilled out of such environments. They are one of the chief ultimate sources of our unwritten axiom, that cities are something to flee from, that the harmonious interaction of man and other organisms can only be achieved out in the country, that the average man is too noisy, too ugly, and too vile to be accepted as a close neighbor.

This native American philosophy is not shared by most of the rest of the world though in parts its roots are North European. In those countries where man has lived the longest, China, Japan, India, the Mediterranean, he has learned to accept himself and to create the kind of city which does not necessarily die at the center, where one can live in quiet and dignified seclusion in the very heart of a thriving community.

The Lesson from Mexico

Some of these countries I have visited in passing, but in Mexico I lived opposite the tiny central square of San Pedro Tlaquepaque, a busy pottery-making and pottery-selling town near Guadalajara. Our front gate opened opposite the arcaded park in the town's center. The wooden doors of our back gate, big enough to let in a five-ton truck, swung back from the main motor-road to San Juan de los Lagos. Between us and these streets, a solid row of small shops: meat market, grocery, bakery, auto-bus depot, separated us from the sidewalks on which they fronted but had no doors into our rooms or patios. Early every morning, through the thick wall which divided our front patio from the meat market, we could hear the heavy thuds as the beef for the day was slaughtered, but if we wished to visit the meat market we had to go out our front gate and back down the street. On the other hand when we spent a warm evening or a sunny winter afternoon on our flat roof our bailiwick there extended right out over the shops. We could sit directly over the bus depot and keep an interested eye on arrivals and departures from the village.

Except on the roof and immediately opposite the front gate, we had complete visual privacy, though we were in the center of a busy town. A few old cabbage palms rose so high from our neighbor's gardens that we could see them in the distance; towards the south our boundary wall was so low for ten or twenty feet of its length that we saw the tops of the orange trees that were in back of the mule depot next door. For all practical purposes when we came in the front gate and closed it behind us we shut out everything else in the village except some of the sounds and an occasional penetrating smell. A few of the noises seemed almost to increase as they got over the high boundary walls, perhaps because so many of the lesser ones were hushed. The cracked bells of the local churches less than a block away sounded in our patio as if they were ringing on our own roof tops, and in the early morning the soft slip-slap of tortilla-making in the back patios of our neighbors seemed to be coming directly from our kitchen.

The ground plan of the house was simple, and essentially Roman. Most of the rooms were small, set back one after another along the boundary walls and opening only on the inner side toward the front or back patio. The *sala*, a large and handsome room, was set cross-wise to the rest of the house, with palladian windows looking out in one direction towards the front patio, in the other towards the rear patio and the small orchard of fruit trees which filled up the corner of the lot.

There we were, in the town—yet out of it. On winter mornings we ate under a big old avocado tree in the back patio, setting up our table in the first spot of sunshine to appear when the bright January sun cleared our eastern boundary wall. Across the patio was a big old laurel tree, its upper half swathed in blossoming Bougainvillea. In the nearer distances were a small banana tree and a large evergreen Magnolia with a brick-lined pool beside it. Beyond were a few orange trees and white-flowered jasmine climbing the walls. The front patio was smaller but more ambitious, though even less trouble to take care of. There were a few tall palms which caught the early morning sunlight when the whole of the town was still in shadow and a row of Casuarina trees whose soft and pendant foliage whispered in the slightest breeze. The canyon wren which made his home in our tile roof, crawled in and out among the tiles, giving his clear ringing song from time to time. Birds came and went in the little orchard trees or among the big palms of the front patio. The groove-billed *ani* (that old stand-by of the cross-word puzzle craze) came in groups of two or three, chasing one another in and out among the palm leaves, more like big black rats with feathers than like any ordinary bird. Little armies of leaf-cutting ants paraded from one patio to the other under their green banners, and bees hummed all day among the orange and lime blossoms.

Yet the village was right at hand, when we wanted it. One step took us to the post office and the bake shop; in the other direction it was less than a block to the main produce market. When I had documents to notarize, the mayor's office was kittycorners across the square. On Saturday nights and during fiestas, like many of our neighbors, we set comfortable chairs just inside the big wrought-iron front gates and sat there protected by the semi-darkness, almost in the holiday crowd and yet out of it.

The big flat roof was a particular joy. There were corners protected from the wind where we worked and ate on sunny winter days. There was another portion where our ancient avocado tree rose like a tent to shade us when the sun was hot. To this day when I open some of my old Mexican note-books and turn the pages I find here and there in the creases, pressed between the pages, the stamens of the avocado flowers which rained softly down during hot days of that Mexican spring. Even more than when we sat by the gate were we, on the rooftop, in the city—yet out of it. The sheltering balustrades were waist to shoulder high and gave us partial privacy. Even though we stood looking down into the square, hardly anyone ever glanced up at us. Here and there across the flat roofs of the town, we occasionally glimpsed another family hanging out washing or making repairs to the roof, but mostly we were alone with the views off to the high hills east of the town, the church towers to the west of us, and the high trees which rose from many of the other patios in town. During the winter we looked down into the bright magenta mass of the Bougainvilleas in our own patio. As spring came on and the *jacarandas* blossomed we could look across the roof tops and admire the soft purple-blue domes of our neighbors' trees. We watched the children playing in the streets, the American lady tourists (their hair blowsy and ill-kept by Mexican standards) buying pottery souvenirs, the doughnut man selling freshly fried crullers from his small portable stove, the poor people from nearby towns starting home with the purchases on their heads, the town drunks being arrested, the arrival and departure of political big shots at the mayor's office. When I had nothing better to do I could deduce in considerable detail the fascinating goings-on of our busy town. Yet by just shifting in my chair I could look down into our patio and watch the bees which came and went among the avocados and the citrus fruits or examine the strange little weedy wild-flowers which burst into bloom after the summer rains were over and grew directly out of the top of the adobe brick wall between our rooftop and the master potter's next door.

Discovering Nature in the American City

I have spent much of my life in a succession of charming old houses, but our Mexican home produced more comfort and more effective relaxation for far less trouble than any of the others. Architecturally it was completely different from our other homes, but its basic difference was in the attitudes back of the architecture. It came out of Roman and Moorish ideas about the privacy of the homes and the acceptance of cities as a way of life. To build a home like it would be illegal in most American towns, particularly if like my Mexican neighbors I kept chickens and stabled burros. It would not fit in our codes and zones; buildings and walls right up to the lot line, the house bordered with little shops, a productive small orchard in the down-town business zone.

There is little hope of my ever living in such a home again. One cannot improve the American city effectively by building Spanish houses. Aping the outward forms of another culture creates more problems than it solves. Changes, to be of any consequence, must come first at a basic, philosophic level. What is needed is not new architecture but new attitudes. Here, as a teacher and writer, I may be of some ultimate effectiveness.

Consequently I have changed my own teaching pattern. I now take my botany classes more frequently to dump heaps and alleys in St. Louis than to the Ozark Woodlands and the beautiful gravel bars of the Mississippi River. We study Trees of Heaven, weed sunflowers in the railroad yards, wild lettuce on a vacant lot. Gradually a few of us are beginning to accept man in our own biological limitations as a real part of nature. The ecology of dump heaps should be more rewarding for the time spent on it than the ecology of grass lands in the Great Plains. In the Plains one must study the interactions of organisms, all of which one only halfway understands. In the dump heap *homo sapiens* is the most overwhelming of all the organisms in his primary and secondary effects on the landscapes under analysis. As gradually we get down to the fundamentals of town and city ecology we may hope to analyze them faster and more effectively because of our own inside knowledge of man and our special insights into what he has been doing. If gradually in this and other ways we can build up a real interest in the ecology of our cities and the fascinating plants and animals which live there with us, we shall have made a very small beginning (but a fundamental and effective one) towards helping Americans live happily in the American city.

Autumn on the Peninsula

Is there no reason why each one of us should not look forward to making a collection of climates as the opportunity presents itself? Summer is always summer, one place as another, and winter is always winter, yet the totality of the experience always varies from place to place, sometimes subtly, sometimes dramatically. The whole pattern of temperature, rainfall and vegetational response weaves its own distinctive landscape in each locale. There are scores of types and sub-types of autumns and winters and springtimes and summers. Only a few are so dramatic that they become written about and pass into general knowledge: autumn in New England, winter in Rome, spring on the Southwestern deserts.

For the collector of climates there is always the joy of discovery. No one has the wit to take the bare facts of day-lengths, temperatures and rainfall, and so skillfully put two and two together that he can predict even the outstanding details of a season in a new climate. So it was that when my professional career brought me back to the Peninsula south of San Francisco, from a September to the following January, I rejoiced in the fact that I was coming for precisely those months which I had never sampled in any previous visits to the West Coast. I could collect a new kind of autumn, not merely a California autumn but one which would in general be representative of all regions with a Mediterranean climate. By keeping a sharp eye on autumn on the Peninsula I would also enlarge my understanding of autumn in the lands of classical antiquity.

The opportunity intrigued me; from books, all that I could predict was that the days would shorten but that there would be no dramatic onset of cold. There would be no hard freezes, at the most only an occasional light frost. I, therefore, expected a more gradual change, compared to autumn in New England or the Middle West, but I was totally unprepared for the attenuated gradualness, a full four months of autumnal experiences, of ripening vegetation, of falling leaves, of brilliant foliage which could be found here or there in the landscape if one sought it out.

For week after week and month after month, the autumnal pageant proceeded, more a series of individual acts than a full chorus. For once, nature seemed to be in no hurry. By the third week in September, the box elders were already beginning to turn a light yellow green; it was not until after Christmas that the wisteria leaves ripened slowly into greenish bronze and fell, a few at a time. The first masses of bright color came in late September with the poison oak which turned to a shiny dark crimson. In the pastures and along the streams it made no more than an occasional splash of brilliant color, but, back in Los Troncos woods, hillsides were red with it; and in the very late afternoon whole landscapes looked almost as bright as in a New England autumn. Before it had gone, the Oregon (or Big Leaf) maples began. Their giant leaves, much like a sugar maple in shape, took on a clear, flawless yellow and so remained week after week. On Skyline Boulevard one would come around a bend and see one tree all in yellow against the background of dark green redwoods. Some of the mountain side-roads wound through considerable stands of them. Seen from a low-flying plane, they were really spectacular where water courses opened out into the valley.

Before the last of the maples had passed, the native walnuts had begun to turn yellow and at Walnut Creek the auto drove through a tunnel of dull greenish gold. The cultivated walnuts of the orchards had already turned a deeper orange, marred here and there by sectors of dark brown. As they ripened they deepened to black and dull orange, then fell in a few days.

A few of our eastern trees made even more of a display than they do at home. The sweet gums in dooryards and gardens were for some weeks a brilliant medley of red and yellow with touches of green and of brown. The gingkoes ripened slowly to a clearer lighter yellow than in the East, just as I had seen them in the Borghese Gardens in Rome, during a previous autumn. In a few gardens there were brilliant trees and bushes I had never seen before. The brightest of all these was the Chinese pistachio; on a neighbor's lawn a whole clump of them began to take on color, scarlet and crimson, the different shades of red interwoven with each other, set off by the shiny green of those leaves and branches which had not yet begun to color up, all the color heightened by the lustrous wax-like upper surface of the leaves.

By late December, all these had gone and only the willows remained. They colored a little at a time until they made masses of golden brown along the creeks and in the Searsville swamp, a color which looked much brighter when there were evergreens in the background for contrast. Finally, on the willows only, the large leaves on the uppermost branches remained, making little pastel daubs of bright yellow brown along the water courses. But by this time the new year had come and brought the first heavy rains. Before the willows dropped their last leaves, acacia buds were just beginning to open and the earliest of the narcissi were in full and fragrant bloom. Autumn in a tenuous way had lingered on to touch hands with a reluctant premature springtime.

Such is autumn on the Peninsula, not a riot of color every way you look, not quick flashes of glory that light up for a few days, or a few weeks at the most, but a long season of "mellow fruitfulness," an experience so subtle that homesick easterners sometimes miss it altogether because it does not force itself upon their attention like the brief autumnal glories of New England. It is, instead, something for the connoisseur, something which, with a little cooperation, may be enjoyed for four full months. One has to learn when and exactly where to look. I have seen one valley all in dull greens and browns, except for Lombardy poplars which rose like a circle of golden exclamation points above a distant ranch house. I have seen the Salinas Valley with pastel yellows and browns in the cottonwoods and willows of the river, though the mountains along its western face were brilliant with wet new snow. I have seen the California buckeyes in hollows among grassy hills, bare of all leafage, their short sinuous branches silvery in the December sunshine, while the valley below them and the hills beyond were purple with heaths and heathers grown for mid-winter flower markets from coast to coast.

To one who has never before passed a full autumn in a Mediterranean climate, there is a more or less unconscious uneasiness at November afternoons which shorten progressively just as at home, but without the great changes in temperature which have always accompanied them before. This and the slow blending of the seasons into one another, the spring flowers venturing into bloom before the dwarf dahlias have opened their last buds, induced a sense that the ways of the world were somehow off the track.

I was spared something of this *malaise* by having spent a few golden days in a previous autumn out-of-doors in Italy and Portugal. These experiences were all of a piece, the general dryness in spite of an occasional rain, the full warm sunshine, the subdued and localized autumn color, set off by glossy evergreens. I remembered in particular a noon-day picnic at the site of Horace's villa in the hills above Tivoli. When lunch was over we lay back on a low hillside covered with rosemary bushes, in full bloom (just as in autumnal gardens on the Peninsula) and noisy with the hum of bees. In the coppice behind us, the native walnut trees were turning dull orange and dropping their leaves. So in California, buoyed up by the conviction that I was experiencing a genuine Mediterranean autumn, I sampled it gustfully on the Peninsula and even bore uncomplainingly the one or two times when a strong wind, gritty with dust, blew all day long under graying skies. So it had for Horace! I was knowing autumn very much as he had known and savored it, and that was quite enough for me.

Reflections on Certain Honduran Gardens

For somewhat over a decade I have been dipping here and there into the technical literature about the origin of agriculture. Some of the pronouncements made me squirm a little, intellectually. They seem bookish and doctrinaire; there is no flavor of the hoe about them. None of them have the air of having been written by a man who had just come back into the house after thinning the radishes or dusting the young cabbages. They distinguish field culture from garden culture; they use precise terms appropriate for European conditions in the last few hundred years.

  As for me, I am not so sure. I keep thinking about my neighbors in the western Mexican town where I lived for six months; I remember the Guatemalan village where I spent a day photographing and mapping the cultivated plants on a few small plots. These experiences bring to mind various things which do not fit neatly into European classifications. Consider, for instance, some of the corn fields and pastures around our little town in Mexico. They were dotted here and there with volunteer guavas and guamuchile trees whose fruit was carefully gathered. Were they orchards or pastures? And the dooryard plants which I have stopped to look at in various parts of the tropics; what words are there in English to describe their groupings? How can one say "garden" when the mass impression was of mingled trees and vines? Will "orchard" be appropriate for an area which is supplying vegetables and cereal crops to a hungry family? I have the uncomfortable feeling that European modern and classical languages stem out of what for the world (seen as a whole in either time or space) is a rather exceptional set of attitudes towards plants and of techniques for dealing with them. We don't even have a simple word for the kind of place which is both garden and orchard; we must resort to a long phrase when we want to talk about all the more or less cultivated vegetation immediately around the house.

If we don't even have the right words to use in writing or talking or thinking, might it not be best to turn our attention first to what people are actually doing with the plants immediately about them? This feeling has led me to keep a sharp eye out in those various parts of the world where my work has taken me. It has finally driven me to watch myself more closely. For exactly what purposes do people grow plants anyway? What are the reasons, all of the reasons, why I, for example, spend so much time taking care of plants? What attitudes towards plants were of most importance in the long process of domestication?

The dooryard orchard-gardens of Central Honduras first came to my attention when I sat in a series of them, studying and measuring the local varieties of maize; they continued to intrigue me for some years after that work was finished. One of my students has begun to look into the problem. In time he and others should find answers to some of the questions which are raised by these preliminary inquiries.

Much of Central Honduras is made up of wave after wave of pine-covered mountain ridges, mostly lightly grazed. Tilled fields—beans, corn and sorghum planted together, sugar cane—are concentrated in the valleys and sprinkled here and there through the uplands. Long stretches of roadway have no houses; in more favorable areas the homes are only a few hundred feet apart. They are mostly of wood or adobe with tiled roofs. Frequently they are nicely made with several rooms and with graceful little porches on more than one side of the house. Close to the house and frequently more or less surrounding it is a compact garden-orchard several hundred square feet in extent. No two of these are exactly alike; the country as a whole is on the dry side and the availability of water greatly affects the composition of the garden. There are the neat plantains (cooking bananas), usually more or less grouped together, much like other bananas but squatter, sturdier, altogether more thrifty looking. There are various fruit trees, frequently the large, handsome "nance" little known outside of Central America, citrus trees of one kind or another, weedy vigorous Melias whose lacy leaves shade the immediate dooryard, a mango here and there, usually a thicket of coffee bushes in the shade of the larger trees. There are tapioca plants of one or two varieties, all the way from one or two plants to several score, grown more or less in rows at the edge of the trees. They are lusty weeds, each plant growing into a good-sized bush when well developed. Frequently there are patches of *Xanthosoma purpurea*, the American relative to the taro, now widely spread in the tropics, a strange plant that has been little studied. It has leaves much like the philodendrons and caladiums we grow for ornament, big jack-in-the-pulpity things of a metallic purplish cast. They are grown for their tuberous rootstocks and in these mountains are so far from their native lowland habitat that they seldom or never flower.

These are the framework of the garden-orchards, the towering nance trees, the fan-like plantains, the clustered coffee and tapioca bushes, framework but no more. Here and there in little rows or patches are corn and beans. Climbing and scrambling over all are vines of various squashes and their relatives; the chayote grown for the green almost translucent little squashes, as well as for its big starchy root; the luffa gourd, the skeleton of its fruits useful for dishrags and bath sponges and various such purposes. It is these cucurbits which give a bower-like air to the whole garden. They clamber over the eaves of the house and may run along the ridgepole. They climb high up into some of the trees or festoon the fence if there be one. Aiding them and abetting them and setting off the whole garden are the flowers and various useful weeds. There may be some small-flowered gladioli such as were common in our gardens toward the close of the 19th Century, there are climbing roses of one kind and another and frequently an asparagus fern or two. There are usually dahlias, little, old-fashioned, fully double sorts of one or two colors, say a pinkish-white and a lavender. There are nearly always cannas, the plain old-fashioned and small-flowered dark red kind.

The Question of Origins

In several of the gardens I saw a weedy form of one of the grain amaranths, of the same general sort that once was fairly common in English gardens under the name of "Love-lies-bleeding." It was one of the great grain crops of the Aztecs and the Mayas and its history and classification have been the subject of a meticulous monograph by Jonathan Sauer. In these gardens it is not ordinarily planted; it is rather a sort of encouraged weed that sows itself. Its plumy red tassels fill in the spaces between the other crops in the summer parts of the garden. The cultivated sorts have white seeds which are not bitter (their weedier, uncultivated relatives have black seeds and a disagreeable taste) and which are still popped and made into little cakes by the Guatemalan Indians. I asked the children at one of these Honduran houses if their family saved the seeds. It turned out that this particular family did not know the seeds could be popped or eaten but that they were saved just the same. They were very valuable, the children said. When anyone in the family caught a cold, the seeds were put in a little bag and hung around the invalid's neck. From Sauer's monograph we know that in pre-Columbian times these grain amaranths were at the very center of elaborate religious rituals. The popped seeds were mixed with sacrificial blood and were molded into little figures which were placed upon the altar. At the conclusion of the ceremony these figures were cut into tiny bits which were distributed to the audience, a sort of pagan version of the Christian eucharist. A few hundred years ago, a few hundred miles away from central Honduras, we know these amaranths to have been one of the great crops of the time and to have figured in many important religious festivals. Is the saving of these seeds as a cold cure a greatly attenuated survival of their former ritualistic importance? One can no more than raise the question from this one isolated observance. It would be interesting to ask a few questions here and there of native gardeners in Central America. We might then have enough data for a reasonable opinion.

Meanwhile the presence of so many ornamental flowers in these garden-orchards has given me something to mull over. These are poor people just at the level of meeting their food needs. Why do they grow so many ornamentals? There is no evident market. These flowers take a good deal of effort in lives which are close to the level of bare subsistence. Are they raised just because they are beautiful plants to look at? Are others, besides the amaranths (dahlias, perhaps, since they were so much used by the Aztecs) treasured for magical purposes? This needs looking into if we are to understand the origin of cultivated plants. In the books on the subject I find almost no words about the origin of garden flowers, yet many of them are very old, quite as old or even older than some of our major crops. I can imagine that some of our oldest domesticates *might* be ornamentals. Primitive peoples along the Tibetan borders still plant ornamental Celosias with bright red or yellow plumes, to scare away devils from grain fields. Until we have exact information on this point it would seem to me quite as possible that man scared away devils with Celosias before he planted grain, as that the grain came first. The beautiful Bixa, sometimes grown in the tropics and sub-tropics as a flowering bush, is more widely used to flavor and color food; even in the temperate zone we import seeds to color butter, cheese, and margarine. By very primitive people it is used as a body paint. Which are older, body paints or agriculture? Woad (*Isatis tinctoria*) persists in cultivation today only in the flower garden; when Caesar conquered Britain our ancestors were painting their bodies with it. Obviously there are a whole series of questions here that we need to look into for a serious understanding of how and when plants came to be domesticated.

Nor are these the only kind of questions raised by these orchard-gardens. European clean-crop agriculture does not transplant well into the tropics. Weeds are hard to control; humus and erosion problems are different there. These continuous production mixtures of trees and vines and vegetables plus encouraged weeds have the look of something which should be carefully studied. We have already learned to use cover crops, to shade coffee and indigo with larger trees. Perhaps these fruit-vegetable mixtures when we thoroughly understand them might even be adapted to mass production methods. They point up a general problem, what seems to me to be *the* fundamental problem of tropical agriculture. As an art, agriculture began in the tropics and very long ago. As a science it began within the last century in Europe and America where it is a much simpler phenomenon both biologically and sociologically. Agricultural science has scarcely begun to dig into the complex biological and sociological problems of the tropics. Those of us trained under the relatively simple conditions of temperate agriculture should go to the tropics not to teach, but to learn.

Living with a Gingko

For the past fifteen years I have been living just across a terrace and a lawn from a hundred-year-old gingko. It is higher than our two story house. For our seven north-facing windows, it dominates the view.

From three upper-floor rooms we look straight into the heart of the tree's branching. It shows much of its criss-cross skeleton when in full leaf. This gives it a special kind of beauty, a beauty born out of disaster, for it is so soft wooded that it loses small branches in violent thunderstorms and nearly every break produces a new cluster of sprouts. Three times in the century it has stood there, the whole top of the tree has been blown out by winds close to a tornado's path. It was not until we became wise in the ways of gingkoes that we recognized these three main rosettes in the present crown. Each was formed by surging new growth that covered the scar and sent out a plume of branches around it, giving rhythmic coherence to the whole pattern.

When there are a few branches on the lawn after a storm, we have learned to welcome them. The whole operation of keeping our lusty old specimen within bounds is like shearing a very high hedge. The shearing is done by the wind; all we do is pick up the branches.

When our gingko needed extra pruning to let in more winter sunshine for the masses of snowdrops at its base, the remedy was simple. No need to call in the tree-pruning crew with their extension ladders and their pruning saws on long poles. We waited for a pleasant day after the leaves had fallen and the framework of the tree was bare. I used only an old iron rake and a step-ladder. High up on the ladder with someone bracing it below, I hooked the rake over a likely branch and gave a sharp pull. Down it came. In about an hour (resetting the ladder took the most time) there was a load of branches for our dump truck. Most of them were the long rope-like ones which hang almost vertically and are closely set with little spurs that bear stamens in early spring. Several, however, were as large around as a man's wrist. It was interesting to find out that one man has about the same branch-pruning ability as the strongest gusts of wind in an average year.

The job was polished off by leaving the step-ladder in place until the noon-time sun demonstrated other small branches shading the snowdrop bed. This was so effective that we left the apparatus until late afternoon and found other branches preventing the afternoon sun and a wide expanse of blue sky (rich in unseen radiation) from bringing winter therapy to our flowers. In late winter the snowdrops bloomed as they never had before and when springtime brought warm enough weather for dining on the terrace, we discovered we could now see nearly all the northwestern sky at sunset.

The greatest delight our old gingko brings us is its display of autumnal gold in those years when the first killing frost comes late. If the gold has time to develop fully, we live in golden light whenever the sun is high in the sky and not obscured by clouds. Though the reflection in through our windows is never strong enough to disturb sensitive eyes, it casts picturesque shadows that set off the yellow in our rooms.

We always hope the first freeze will come on one of those typical Indian summer mornings with clear, cold, quiet air. Then the leaves begin to fall gently like big snowflakes, shortly after sunrise. At first they fall one at a time, then come faster and faster, forming by noon-time a circular carpet that is a true reflection of the branches' outreach. The first day this is a golden carpet. With age it darkens to a duller yellow like the true "imperial gold" of some Chinese rugs. For practical and esthetic reasons we always hope that the leaves will fall on a quiet morning. Under those conditions much of the leafy cover is above the snowdrops. It makes an effective mulch. The rest are on our little lawn and can be raked up quickly. Last year the first freeze was a violent and early cold wave with strong winds. There was no rug. The leaves blew every which way in unsightly confusion, yellow-brown litter that had to be swept up and raked up and carried away.

This gingko is an almost perfect screen for a home in one of the most public areas in the Botanical Garden. From late spring to autumn it gives visual privacy from the streams of visitors who pass so close to our dining terrace. It does this better than would another kind of tree, for its changes are almost perfectly timed to our needs. When colder air and lengthening shadows make our terrace too cold for dining, we welcome the dramatic change that from all our north windows gives us long vistas through the Botanical Garden. In the spring when we are in no hurry, neither is the tree; changes come slowly. It is a greenish veil through which we glimpse masses of early flowering trees and shrubs in the middle and far distances. At just about the time these blossoms go and it is warm enough to use the terrace, our green gingko curtain is in place for the season.

Old gingkoes are practical for parks, campuses, and big estates. They make impressive and unusual backgrounds at minimum expense. They have no pests. They date back to the dinosaurs and seem to have outlived the insects and diseases that once preyed upon them. They are smoke tolerant and withstand salt in the subsoil. We need to develop in this country the kind of reverence for picturesque old ones common in various parts of the Mediterranean Basin and in the Orient.

Old gingkoes have two faults, but they would cancel each other out with proper management and intelligent planning: they do not relish the shade of other trees, even the shade of another gingko; and the plum-like nuts of the female trees are covered with a sticky pulp which smells like rancid cheese for some months. It is possible, but difficult, to grow cuttings from male trees that will bear no fruit. They are expensive and frequently are not easy to get.

Young gingkoes of either sex make attractive and interesting borders along a path on a campus or in a park. It will be some years before they are mature enough to produce flowers. During this time one would need to build up interest and understanding of the long-range program I am about to suggest. It is based on living next to our ancient gingko and watching many others (I began gingko watching at age ten).

When your row of gingkoes reaches adolescence pull out every female tree as soon as it bears its "silver apricots," to use the old Chinese name for them. *Do not plant any tree in its place.* Its two neighbors, whatever their sex, will already appreciate the extra sunlight. The result within a decade or so will be a row of increasingly picturesque gingkoes.

The row will be made up of trees *at irregular intervals.* This may be distressing to gardeners with neat minds but "irregularity is opportunity" in designing gardens, locating highways or planning for new buildings on college campuses. These old-gentlemen gingkoes are as adaptable as they are picturesque. They can be more charming at 200 years than they were at 100. They are treasured in Japan. Will American planners and executives permit us to see one in this country?

The Country in the City

Most of my professional life has been spent in botanical gardens, islands of greenery in the midst of big cities. Both at the Arnold Arboretum in Boston and at Shaw's Garden in St. Louis, I have been increasingly fascinated by the peculiar problems (horticultural, biological, sociological) posed when, year by year, areas of quiet rural charm become buried deeper in a close-packed city. So far as I know, the fundamentals of these problems have been given little direct study; apparently no book or conference has attempted to set forth the difficulties of displaying the countryside, or a reasonable imitation of it, in the midst of a city. Yet the impasse grows graver and more pressing with each decade. It is not merely that now there are more and bigger cities. It is that the city now penetrates the country. With our technological progress, the capacity of cities to urbanize the surrounding countryside is making our entire eastern seaboard an intermeshed web of metropolitan influence.

When in 1919 I first went to the Arnold Arboretum as a student, the "cityness" of Boston stopped at the city's edge or nearly so. The back corners of suburban Brookline and Needham were then almost as truly rural as if they had been appended to villages in central Massachusetts. For the Arboretum itself, the difficulties of maintaining its brook-centered meadow, its oak groves, its rocky hill of hemlocks, its marshland with wide sweeps of goldenrod and joe-pye weed, were just beginning to be felt. Now the brook is dry much of the year, some of the oak groves are gone, the marshland no longer looks like an old-fashioned swampy pasture. Today the forces responsible for these changes have full play not only within metropolitan Boston; some of them would be perceptible at almost any site between Boston and Providence, so completely has the city spun out along the highways.

The peculiar problems raised by attempting to preserve country landscapes in the midst of cities are largely unexplored, even on the technical side. If you would like to come face to face with an important one, pick out a flourishing woodland in upstate New York or central Vermont and seat yourself on the forest floor. All around you is the brown thatch of fallen leaves. Choose a space about as big as a card table and pick away, one by one, the leaves which fell last autumn. The surface of the previous year's leaf fall is revealed, more or less compacted by rain and weathering. With a little patience this too can be removed, sometimes more or less as a continuous sheet, and then another layer is revealed beneath it and another beneath that. When, however, one has peeled off three or four of these successive autumnal wraps, the character of the underlying leaf deposit changes. There are spots where insects have been at work, the delicate hyphae of molds and other fungi make fans and circles of thin white webbing as they grow over the leaf surfaces. With such digestive processes under way, the leaves themselves become fragile. First the soft tissue between the veins disappears, then the lacy skeleton of veins breaks up. At about this point, it becomes increasingly difficult to peel off the layers. It is not merely that the leaves are brittle and disintegrating; tiny rootlets are growing in among them; they are actually stitched together with little roots.

Now take a sharp knife and cut down into the leaf mold, bringing out a cube of material the size of your fist or a bit smaller. Shake it gently to dislodge as many of the fine particles as possible; swish it carefully back and forth through the water if there is brook near by. What is left in your hand? A red-brown snarl of crooked, wiry, knobby rootlets. If it is from a rich and healthy woodland, there may be so many roots so close together that, even after you have washed off most of the soil, you can still see quite clearly the outlines of the square-faced cube cut out by your knife. You will find just as many of them if you take your sample from the bare forest floor between the trees as if you dig close to a bush. Where do all the rootlets come from? Why, from the very trees whose branches spread out overhead. Examine them closely; they are no ordinary roots. They are not uniform in diameter. The little knobby areas all along them are specialized feeding zones where, with the help of fungi, the nutriment in the dead leaves is digested and sent coursing down into the bigger roots and then up into the tree. These are roots that grow upward, up from the larger roots into the zone of rich leaf mold which accumulates year after year on the forest floor. This brown carpet of leaves on which you are walking is one of the main feeding zones (not the only one to be sure, but an important one) of the big forest trees all around you. Here you have one of the fundamental difficulties in maintaining woodlands of this character in or near a city. The kind of trees which grow in rich upland woods are accustomed to a layer of forest duff around them and they will not be happy without it. It is not merely that they have had it all their lives. The parent trees from which they came were also born to it and *their* parents before them. They have evolved in woodlands with thick blankets of leaf mold, and if one is going to grow them well he must either provide the mold or its equivalent. Food is not the only purpose it serves. It protects the ground from erosion, soaks up rain so that it does not run off, and prevents rapid evaporation in dry, windy weather.

Maintaining the leaf mold on a forest floor in a truly urban area is impossible or virtually so; this leaf carpet is not so tough a rug as it looks. Even with only a few visitors each day, it begins to wear away. In a fine tract of ancient woods recently acquired by a big university, I discovered that the mere passing back and forth of the students and teachers who were attempting to study the woodland and analyze its basic ecological problems had worn the duff away in the areas close to the recording instruments which they had set up. Fortunately man seems to be by instinct a path-following animal. If one makes paths through such a woodland, most of the visitors stick to the paths and the forest carpet gets less damage. Note what happens, however, in any rich woodland where a family with a strong bent towards natural history acquires a camp. If they picnic in the woodland to any extent, the carpet is worn away, the feeding rootlets are injured or destroyed, and within a few years the big trees (deprived of their vitamin pills) start dying at the top. If there are dogwoods present, they blossom less and less each year.

Nor is the wearing away of this leaf carpet its only danger in urban areas. In dry spells it becomes inflammable and a severe fire hazard; protecting it in an urban area becomes almost impossible. With artificial fertilizers, with various mulches, we attempt to replace it, but the technical problems of keeping the mulch-feeding zone alive and healthy are difficult ones, in even semiurban areas.

Luckily there are other kinds of woodlands. In flood plains the river carries away the leaf deposits; trees bred to fit in with the vagaries of rivers—silver maples, sycamores, pin oaks—take more kindly to urban conditions than do their relatives which grow in rich upland woodlands, sugar maples and white oaks, for instance. Likewise, the trees of dry ridges and rocky hills have been bred to little or no association with leaf mold, but these areas present other and equally difficult problems under urban conditions.

Loss or deterioration of the leaf mold is only one of many technical facts which must be faced, if one is to keep something of the country healthy and attractive in the city. The total effect of the urban environment is much stronger for some organisms than for others. The red-bellied woodpecker is seldom seen in Shaw's Garden in St. Louis, though in the outer suburbs it is almost as common as the hairy woodpecker, and the latter does come into the city. The sparrow hawks visit us frequently, but not the red-tailed hawks nor the marsh hawk. The charming little European tree sparrow has followed the horse barns out to the very edge of the city.

A few organisms are so supersensitive that they stay completely outside the metropolitan zone. Others spread way into the city, but do not reproduce themselves abundantly as they do out in the country. Only on the city's outermost fringe does our native red cedar vigorously seed into nearby pastures and rough lands. Within the suburban zone it grows well and gives rise here and there to occasional seedlings. Well within the city it can persist if transplanted, but no seedlings are found. The old persimmons in our botanical garden are apparently in good health and some of them fruit heavily every year, but persimmon seedlings come up no longer in waste places within the city limits. Introduced plants may behave in a similar fashion. The Korean lespedeza which mantles a good part of the Ozarks (even where it has not been planted) stops almost at the city's edge.

The most important effect of this differential tolerance of urban conditions is an indirect one. The organisms which can survive are thrown into a new set of relations with each other. The complex web of life is out of balance. For the plants and animals which can tolerate the city effectively, there is less competition than before, and some of their predators are no longer at hand to keep them in check. Squirrels become a real problem in our city garden. There are no hawks to hunt them, no coons to compete with them for food. They increase so rapidly that a house in the grounds may have to fend off five to ten of them every fall when they are looking for nesting sites. They eat all the berries off our holly trees in the more rustic areas of the garden and claim the lion's share of the food put out for the birds in snowy weather.

Most challenging of all is the effect of these changes upon the naturalist himself. Few indeed are those who are adjusted to it or have attempted to think through a rational nature ethic for the city. Most dedicated naturalists are emotionally attached to nature as they find it in country meadows and woodlands. They would like to bring it, in this same well-established balance, right into the city. When this attempt fails, they go out of balance too. The same retired school teacher who spends her summer vacations feeding prune pits to the ground squirrels around her mountain cabin will refer to the squirrels around her old home in the city as "nothing but rats with bushy tails" and set traps for them with grim determination. A group of ornithologists or other naturalists will agree reasonably well as to general policies in maintaining a wild-life refuge in the country. Let them try one in the city and no two of them will be in accord as to what should be done about squirrels, English sparrows, blue jays, and cats. Perhaps they will all agree that the cats should be killed or kept out somehow; but should the blue jays be encouraged, or shot, or be kept away from some or all of the feeding trays by the use of appropriate devices?

Two naturalists of my acquaintance had become firm friends during days spent together in the field. Then one of them began birdbanding in suburban St. Louis, maintaining traps under the bushes in his garden. When cardinals or chickadees (or even robins) came to the trap, they were provided with a numbered band and set free again. When English sparrows ventured in, they were promptly disposed of. One Sunday the operator of the trap entertained my other friend for dinner and the latter was pressed into helping put on the little aluminum bands and making the records. He arrived at my house a few hours later so shaken by the experience that he had to talk it over with someone. "The first part of it," said he, "was wonderful. I've never known anything like it before. The little chickadees and juncos would lie back quietly in your hand once you had hold of them the proper way. They didn't seem to be disturbed by having the band put on. I had never realized quite how small they are and how finely made. The little things would lie there on your hand after you'd opened it up, then they'd flick their wings and be off into the air where you couldn't follow. But those poor sparrows! It makes me sick. I had no idea he was that kind of a guy. It wasn't just that he killed them; it was the way he did it. He hates them. He hates them so fiercely that he gets some kind of a kick out of killing them. There they were, these defenseless little birds that had come into his trap. We took them down to the furnace room and opened the furnace door. Then he took the sparrows one at a time, wrung their necks and tossed them in on top of the coals. *And he was happy doing it.* You should have seen the glee on his face when he wrung the necks. I don't think I can ever go birding with him again with any pleasure."

This sudden break between two old friends comes closer to the basic difficulty of displaying country landscapes under urban conditions than one might think at first sight. Naturalists who will not face resolutely the fact that man is a part of nature cannot become integrated human beings. A nature-study movement which focuses its attention on remote mountains and desolate sea marshes is making a sick society sicker and sicker. The ecology of cities could be a fascinating study for the amateur or the professional biologist. Nature may be out of balance, but it is when phenomena get manifestly out of balance that one can most easily analyze the interplay of the underlying forces. If more naturalists would accept homo sapiens they would turn their attention more and more to the plants and animals with which he spends so much of his life: trees of heaven, squirrels, sunflowers, dogs, dandelions, cats, crab grass, English sparrows, gingkoes, weeds of all sorts. We would in time learn the dynamics of waste lots in the city, of dump heaps, and of city parks. We would know what is and is not practicable in bringing country values into city landscapes. More importantly, we could acquire a fellow feeling for these organisms with which we live. We would accept cities instead of trying to run away from them, and in accepting them, mold them into the kind of communities in which a gregarious animal like man can be increasingly effective.

Islands of Tension

If you are on the staff of a botanical garden or arboretum you never know when you answer the telephone what the call may lead to. With no warning a phone call one April afternoon began, "Dr. Anderson, this is the War Department calling. Can you report at the government wharf in South Boston at eight o'clock tomorrow morning for the committee's official visit to the islands in Boston Harbor?" Most of these islands were under the control of the War Department, the Coast Guard or the City of Boston and joint tours of inspection were made from time to time. Since erosion was becoming a serious problem, a committee had been set up to study it and to inspect tree plantings made some years before. As a staff member of the Arnold Arboretum I was one of the experts added to the group.

Since I first saw these islands, fifteen years before, they had fascinated me. I'd ridden repeatedly on all the ferries or excursion boats which then plied across the harbor and climbed all those promontories along its margin which were accessible to the public. When I became a member of the Harvard faculty I looked into the possibility of visiting such spots as Governor's Island, but gave up the idea when I learned that one needed the blessing of the War Department.

The first week in April is not the best time for planning a visit to Boston Harbor. The snow may be gone but the air is raw. Lawns are just beginning to green up; nothing much is in flower but pussy willows. The next morning brought us all that Boston can hope for at that time of year. Though cool, with a steady breeze, it was cloudless all day, pleasantly warm wherever you could get in the sun and out of the wind.

It was a mixed group of about thirty men who met at the wharf. A few of them were well informed about the islands. One of these, Patrick J. Connelly, president of the Dorchester Board of Trade, was an authority on the islands and their complex histories. He had recently published an attractive pamphlet, *Islands of Boston Harbor, 1639-1932, Green Isles of Romance.*

After winding in and out among the islands we landed on the largest, Long Island, to inspect tree plantings made in about 1910. They had not been well cared for and the choice of trees had evidently been made without technical advice. They were common European species, easy to grow in nurseries but not the most promising things for bare little islands swept by cold winter winds and salt spray. Some trees had died. Those that remained were English oaks, European white birches, Scots pines and Austrian pines, of which only the latter were in fair health.

Although farther out in the Atlantic, other trials on Gallop's Island looked more promising. As the most prominent island in the outer harbor it has been a quarantine station since pre-revolutionary times. A doctor at the Quarantine Hospital had been trying out likely trees and shrubs since about 1927 and some of these seemed to be doing well: Manchurian ash, Carolina poplar, privet, sorbaria and Amur cork tree.

I was disappointed that apparently nowhere in the harbor had the Japanese seaside pine, *Pinus thunbergii*, been given a trial. By the time of this harbor tour it was beginning to look promising at exposed oceanside locations in southern New England. Since then it has done spectacularly well at Jones Beach and its peculiar merits are widely known along the East Coast.

Long Island had been a kind of dumping ground for the poor of Boston since 1885 and the plantings we inspected were near a cluster of hospitals, administrative buildings and a fine new recreation center on a high bluff at its northern end. The schedule called for a tour of the whole island. Two of the officers led a half dozen of the more earnest and vigorous members by narrow paths along the low cliffs above the beaches.

This route gave us an almost continuous view of the foreshore. I was immediately struck by the great number of orange crates and unsightly rubbish in the zone of driftwood. Immediately above the orange crates were occasional low rosettes of an unusual rose, one of the Arboretum's Oriental introductions with which I was familiar, *Rosa rugosa* var. *kamtchatica*. It differs from the ordinary rugosas of our gardens by being generally smaller with a more spreading habit of growth. It had certainly not been planted there intentionally but was already of some importance in lessening erosion on the upper margin of the foreshores. Its buoyant orange-red fruits had put down roots where they had been cast up by the high waves of winter storms. A month later I saw more of them along the magnificent beaches on the outer arm of Cape Cod, as well as a single specimen of the ordinary bushy *Rosa rugosa*. From the technical literature I learn that the Kamtchatca rose was originally native to the same upper beach zone in the northern Orient. Now and then from an intensive flora of some New England island or estuary, I have learned that it is apparently still spreading along the northern coast of New England.

These scattered bits of information have more significance now that the whole problem of evolution on beaches is being rigorously and comprehensively studied by my former student J. D. Sauer, jointly Professor of Botany and Professor of Geography at the University of Wisconsin. In his world-wide analysis of tropical beach vegetation he is demonstrating that life on beaches is so rigorous that precious few species of the world's flora can persist there. The few that can take it have little competition so, in general, on sea beaches many individuals of a few species are spread over wide areas. Furthermore, now that Sauer has pursued these studies in both the Old World and the New, he is demonstrating for an increasing number of beach plants that when they find their way from one hemisphere to the other they fit into the same kinds of situations in their new home as they left in the old.

Though it was the general problem of eroding sea cliffs and foreshores that had taken me to these islands, as our trip continued I became more and more impressed by the sociologically specialized environments of the human communities which shared these islands. The local and national needs the islands served fell in a few widely diverse categories. They regulated and protected maritime traffic. They served to isolate contagious diseases and social misfits. Their potential for recreation had been realized only at Castle Island and even there only when its ancient forts became obsolete (as well as increasingly picturesque) and filled land connected the island effectively with city parkways.

On Long Island we came upon a beach with a protecting cliff above it, where some of the inmates had built themselves little "clubhouses" out of driftwood and other scraps. They varied from crude hovels to weathertight structures with chimneys and windows. What other reactions to the harbor's peculiar environments would one find if he made a real study of the whole problem? There is a little to be gleaned from Mr. Connelly's eloquent compilation. Those living on these islands were under increasing and varied stringencies in the three hundred years covered by the booklet. The lighthouses, the quarantine stations, the military installations, the public parks, the institutions for unfortunates, the garbage disposal plants, had not only taken increasing space, they operated through different offices. It is bad enough to have your fate in the hands of a government bureau; it is worse to have it decided by bureaus which may be at odds with each other and whose certainty of public support varies with the times.

One of the changes I wonder about is the effect of mass-produced pleasure boats of all kinds. On other shores I have witnessed their increasing effects not only upon human existence along water-fronts but the chemical and biological changes they bring to the beaches and the very water itself, as well as to the plant and animal communities within and near it. Since World War II speed boats must have brought complex problems to Boston Harbor and its islands.

The overall effect of such various and shifting pressures on human existence is even more violent than the stringencies reported for the plant communities of sea beaches by J. D. Sauer. The urgent and conflicting demands of national defense, protection and control of maritime traffic, waste disposal, recreational needs of a crowded city, isolation of contagious diseases and of social misfits are reflected in the human population of the islands.

A few details are reported in Connelly's booklet: after the old fort on Governor's Island had been abandoned a squatter made his home in the ruins and his body was found there after his death. During King Philip's War (1675-1676) whole villages of captured Indians were confined on Deer Island and hundreds died there from starvation and exposure. The boys' reform school on Rainsford Island was abandoned after the boys cornered the Keeper down on the beach and stoned him to death. During the Civil War a whole group of Southern generals were confined in the military prison on George's Island. For many years a hermit lived in a hut on the southern shore of Slate Island. Before modern hospitals were available, Bostonians ill with contagious diseases were buried in the little cemetery near the Quarantine Hospital.

Even the plant communities reflect the violence of these various tensions. Just as ordinary sea-beaches are limited to many individuals of a few species, so on these islands they may be restricted to even fewer. At the time of our visit, Governor's Island was covered by the most rampant thickets of poison ivy I have ever seen. The watchman's dog had died from repeated exposure to it. It seemed to be growing in practically pure stands. On two islands where summer homes or hospitals had been abandoned there were thickets of Staghorn Sumac. These were not accompanied by other woody plants as in ordinary beachside communities, but were solid masses of sumac.

Before our tour the islands of Boston Harbor had appealed to me, to use the phrase from Mr. Connelly's title, as "Isles of Romance." Since that day I have increasingly come to think of them as islands of tension, tensions so violent and so various that their interactions might profitably be studied in some detail.

The City Watcher

Over much of the world man has been accepted as a city-loving (or at the very least a town-loving) creature. Where this philosophy is underlying, the city-dweller can easily be at one with nature; the sensitive Zen Buddhist, even though poor and city bound, may expect to derive those values from the landscape around him that he might in a small village. Over much of the world, man has a courtyard, with seclusion if not quiet; from the Mediterranean lands eastwards he may hope for a rooftop.

    This is all very well in other parts of the world; what may the intelligent American do to temper his present dilemma; forced as he is to choose between (1) increasingly hectic daily traffic-rushes to increasingly less rural suburbs or (2) to reside among increasingly unpleasant surroundings in increasingly decaying city centers? Not merely what are we *all* to do about facing this general problem philosophically and intelligently but how is Mr. John Smith of 4756-A Washington Avenue to act when his lease runs out the first of next September?

    For a few short months I had to make this choice. I solved it so easily that I now look almost nostalgically to my days in the Saum Hotel on Grand Avenue, particularly my early winter mornings and late spring evenings. We had only one room, with a Murphy bed, an absolutely minimum bath, and an alcove kitchenette. It was merely on the fifth floor but we had a big west window and the hotel being on a slight ridge we were well above all the neighboring rooftops. By using two low window boxes to screen the base of the window we had complete privacy from the outside world; the curtain was left up day and night and we lived with the sky, the wide undulating stretches of rooftops, with country hills in the far distance. It soon became so fascinating that we moved our dining table up sidewise against the window boxes and sat down to breakfast and dinner facing the view. For long stretches of the winter we ate breakfast with fog or haze down below us; sometimes so dense that the city was shut out of the picture and the rising sun brightened the occasional steeple or tall building reaching up to share with us the startling clarity of the upper air. No two days were alike. I remember vividly one morning when we were out in the clear. The Georgian dome of the City Infirmary in the middle distance rose above the all-blanketing fog. Another morning only the tip of the dome could be clearly seen and off to one side the black outline of a water tank.

Just as Thoreau, moving to Walden, became increasingly conscious of the risings and fallings of Walden and Fairhaven Ponds, the dates the pond froze, and its crackings and boomings on winter nights, so we from our high perch, took for the first time an intense day-to-day interest in the fog and haze, the smogs and storm clouds, the ever-changing air around us. I look back with particular pleasure to a day of little showers in early summer. My wife was out for the afternoon and I was alone with the sky, working with my papers spread along the table. First there were in the bright blue sky, little puffs of clouds here and there, like meringues on a custard. Gradually they grew and multiplied. Every time I glanced up from my work there were fewer and fewer spots of sunshine about the city. I remember with delight seeing a big white apartment house in the edge of the distant suburbs suddenly flash into view when highlighted by one single shaft of brilliant sunshine, then fade quickly into the background. An hour later rain was falling here and there about the city. I stopped from my work to watch several little showers take their individual slantwise paths from northwest to southeast, much as on a summer afternoon in the high plains of Texas.

"This is all very well for you," someone is sure to say. "How is the average busy man to find quiet moments for meditation and refreshment? We cannot all have a big window that looks out over the city." To which I reply that in the first place it depends on his sense of relative values. My hotel was not the smartest in the city. It was clean and decent but in the German-American part of town, as unknown to the would-be intelligentsia of our city as to the suburban carriage trade. But if one accepts Man as a part of Nature there is always something to be found. Your pseudo-Thoreaus are quite at ease sitting on a hilltop in Dublin, New Hampshire, watching a few farmers at work in the distance; or picnicking near a Maine wharf, observing the fascinatingly quaint lobster fishermen in their comings and goings. For the past five years my professional duties have taken me to confining, day-long meetings at the National Science Foundation headquarters near the White House in Washington. Frequently it is too cold and rainy to go out and sit on one of Mr. Baruch's benches during my few periods of respite, so I duck into the little old colonial church across the street and settle myself down in one of the back pews. Sometimes (though rarely) it is really quiet and I, being a Quaker, settle down to silent meditation in good Quaker fashion. Usually there are all sorts of comings and goings, and I find Episcopalians and their visitors and friends quite as suitable and interesting objects for nature study as are Maine lobstermen or New Hampshire farmers (or even, for that matter, chipping sparrows and bluebirds). They have colorful and apparently purposeful and satisfying folkways. As a *sympathetic* observer I sit and watch them, easing my mind for a few moments from the difficult problems of how to encourage basic research with government millions without at the same time stifling initiative and originality.

I must say that I have rested and meditated and observed in a good many churches in this fashion and have learned a lot. I am always puzzled that more of my anthropological friends do not try this avenue of approach in understanding the various cultures they visit. I have sat in the back corner of country and city Catholic churches in Mexico and Colombia and Honduras, in New York City and in central Iowa, watching with a trained eye and an increasingly understanding mind. I am closer to Catholics than to chipping sparrows and can apprehend much more by observation alone. I find there are many things my Catholic friends in St. Louis know little about, such as the contrast between the crowded churches of Medellin in Antioquian Colombia and those over much of Mexico: the keen forthright clerics in Medellin, the beautiful music, the intelligent discourse in clear Spanish, expertly amplified to every corner of the edifice; the throng of worshippers representing each class in the city; nearly as many men as women. In Mexico the church is often old and charming. A few poor people come and go quietly, lighting candles in chapels and praying long and earnestly on their knees, now and then well-to-do women, conspicuously over-dressed, accompanied sometimes by little boys in Eton jackets and white starched Eton collars, make a great show of telling their handsome rosaries. Far up front at the altar a priest mumbles some kind of service in a quick murmur so low it does not tell me whether it is Latin or Spanish or a mixture of the two. In Honduras and in New York and in the South German and Bohemian farming centers of central Iowa there are still other things to watch in tranquility, to turn over in the mind. One can forget one's own troubles, and find peace and quiet, and food for thought in the intelligent observation of nature. It is quite as easy in the city as in the country; all one has to do is accept Man as a part of Nature. Remember that as a specimen of *Homo sapiens* you are far and away most likely to find that species an effective guide to deeper understanding of natural history.